Science and Morality

Science and Morality

New Directions in Bioethics

Edited by

Doris Teichler-Zallen
Nazareth College of Rochester

Colleen D. Clements
University of Rochester
School of Medicine

LexingtonBooks
D.C. Heath and Company
Lexington, Massachusetts
Toronto

This book is based on papers presented at three conferences on science and morality organized by and held at Nazareth College of Rochester, Rochester, New York.

Library of Congress Cataloging in Publication Data

Main entry under title:

Science and morality.

Papers from 3 annual conferences beginning Spring 1979 and sponsored by Nazareth College of Rochester.
1. Medical ethics–Congresses. 2. Bioethics–Congresses.
I. Teichler-Zallen, Doris. II. Clements, Colleen D. III. Nazareth College of Rochester.
R724.S394 174′.2 80–8926
ISBN 0–669–04406–7 AACR2
ISBN 0–669–09808–6 (pbk.)

Second printing, November 1984

Published simultaneously in Canada

Printed in the United States of America on acid-free paper

International Standard Book Number: 0–669–04406–7 Casebound

International Standard Book Number: 0–669–09808–6 Paperbound

Library of Congress Catalog Card Number: 80–8926

Contents

List of Figures

List of Tables

Preface

Scientific developments of the last thirty years, and especially the last ten, have had a tremendous impact within the scientific community and have raised major ethical, moral, legal, and political questions for the whole of society. These questions have arisen in basic research areas and in actual clinical practice. Recombinant DNA is the prototype of an area of basic research in current biological science that is a focus of such concern. Ethical aspects have always accompanied the healing arts of clinical medicine, an activity that is an applied science rather than a basic one. Recent advances in medical research, however, such as in vitro fertilization and techniques for prolonging human life, have particularly heightened the ethical issues. Yet despite the very great need, few mechanisms currently exist for a creative interaction involving the various groups whose participation would be crucial in meeting these new problems and challenges.

To provide a forum for such an interaction, Nazareth College of Rochester instituted a unique series of conferences with the title: Science and Morality—A Dialogue between Scientists and Moralists. The first conference was held in the spring of 1979, and two others have taken place annually since then. In designing the programs, considerable emphasis has been given to establishing communication between the disciplines and to achieving a balanced interdisciplinary approach. The scientist participants were contributors to their fields who had also previously addressed ethical issues. The ethicists (philosophers and theologians) were already familiar with scientific literature and terminology. This assured that actual dialogue and relevant discussion could take place. Attention to accurate assessment of the scientific and clinical foundations also allowed the unrealistic outcomes, or scenarios that are more science fiction than science and that only create confusion, to be ruled out. The audience was composed of individuals of diverse backgrounds and interests and included members of academic and medical school communities, theologians, lawyers, medical practitioners, and the general public. Such a melting pot of interests and expertise served to encourage open and informed discussion.

The collected papers from the three science and morality conferences form the chapters of this book and have been combined and divided into three parts. The introduction serves as an overview of the field and the parts that follow focus on issues relevant to a specific phase of the life cycle. Part I deals with those issues involving reproductive choices and genetic knowledge, such as genetic counseling, in vitro fertilization, and recombinant DNA technology. Part II is concerned with chemical and behavioral modification and environmental issues. Part III is devoted to euthanasia, suicide, and hospice-movement issues. Introductory comments preceding each part have been included to provide a background in scientific terminology and ethical analysis that we hope will be helpful to nonspecialists in these areas. These comments represent the judgment of the editors, who are solely responsible for any errors or inadequacies contained herein.

We are grateful to the faculty, staff, and students of the Nazareth College community whose commitment and cooperation have made the conference series so successful. President Robert A. Kidera has been a continuing source of encouragement and support. Sr. Marion Hoctor, Robert McCambridge, Richard Matzek, and Conference Secretary Barbara Dinse have provided invaluable aid in planning and conducting the conference programs. Grants from the Gannett Foundation and the New York Council for the Humanities have enlarged the conference activities and allowed all the sessions to be open to the public at no charge.

These conferences and this book would not have been possible without the efforts of the Conference Organizing Committee: in addition to the editors of this book, Sr. Therese Lang (chairman, Chemistry Department) and William Shannon (chairman, Religious Studies Department). These two colleagues shared in structuring the conference issues, selecting the best participants to deal with these issues, and making this discussion of bioethics easily accessible to the community. Their hard work, enthusiasm, and creativity contributed greatly to the success of the conferences.

Special appreciation is due our colleagues who were generous with their help in the preparation of this collection—Richard Doherty and Richard Zallen for their important comments on content and organization, Sr. Magdalen LaRow for preparation of illustrations, Sr. Margaret Teresa Kelley for editorial advice, Jay DiGaspari for typing and correcting the manuscript, and Darcy Clements and Irmhild Zimmerman for secretarial assistance.

The Nazareth Conferences on Science and Morality will continue, bringing together scientists, clinicians, and ethicists who make an attempt to articulate the values of science and to clarify human purposes and choices with the goal President Kidera stated when he opened the first conference in March 1979—that we all "might know more clearly, understand more fully, and decide more wisely."

Introduction: The Direction of Bioethics

We are in the second generation of bioethics, as one of the pioneers in the field observed, and the second generation always has an identity problem. "Science and Morality" is the identity problem. It is a way of describing the field of bioethics—the combination of the life sciences and ethics—but the goals and methodology of this combination need to be made clear. What seems like an easy coupling of words is not as understandable as the first generation of bioethicists' thought. Is bioethics a specialty within ethics, or within science, or both? Is it primarily concerned with developing ethical tools? With, for example, practicing good clinical medicine? Or with developing some mixture of both skills? And to what purpose? This last question is a crucial one for bioethics, but its answer is not at all obvious.

We intend *science and morality* to be an interdisciplinary statement primarily concerned with scientific value systems and the development of a modern ethical theory to articulate those systems, and always with the goal of benefiting individual human beings. There are alternatives. We could see *science and morality* as a statement about social goods or concerns, the apportionment of resources, the assessment of collective risks, the realities and ideals of power. We could regard it as a theory about the search for a just order or the recognition of principles. We could see it as responsible for refining terminology and being clear about word usage.

We leave to the reader, and the future development of bioethics, the task of determining which of these alternative meanings for this statement can be the most relevant and productive. The collected chapters in this book may vary in their emphases, but to some extent they represent a choice: that science and morality is a statement about unity, not dichotomy. In medical ethics, for example, the purpose or goal is the patient's well-being and thus represents an expression of the value system of medical practice. In environmental ethics, the goal is the use of scientific technology for the welfare of individual citizens and an expression of the value system of technology.

Bioethics, then, is interdisciplinary, but there are different approaches to achieving the desired interrelationship between science and morality. In one approach, an established tradition or discipline such as ethics, with its own assumptions and methods, examines biology or biochemistry or medicine; identifies what, from its viewpoint, it considers to be major problems; and calls into question certain aspects or applications of the scientific or medical fields. The scrutiny proceeds in only one direction. The corpus of science or medicine critiqued is not permitted to call into question any aspects of the field of ethics. Much of the current literature assumes this one-way form, and it is often provided by professional humanists and ethicists with little or no hands-on experience in science or medicine.

In a second approach, individuals who perceive internal conflicts and contradictions in science or medicine use tools from other disciplines (for example, ethics, theology, sociology, and so on) to advocate a radical change in the scientific system. They have already identified what they feel are problems, already have a programmatic change as their goal, and attempt to validate this goal by the use of another field's assumptions and methods. The scientific area being critiqued has already been judged to be in need of radical change, and the assumptions and methods of the discipline used to justify the change are, once again, unquestioned. This approach is often used by professional scientists or physicians who have a knowledge of the second discipline. Like the first approach, this one is essentially unilateral in its stance.

A third alternative, and one that we feel is the most fruitful, resembles the great interdisciplinary sciences (for example, biochemistry, biophysics, and molecular genetics) in that the assumptions and methods of one discipline are not imposed on the other but each field is, instead, open to question and modification by the other. The direction of change (and, we hope, progress) goes both ways. Neither field's foundations emerge unchanged as a result of the union. Workers in such an interdiscipline need to adequately understand both fields, if not specifically and in a highly specialized way, then certainly foundationally. It is not necessary, for example, to know every citation in a review article on utilitarian theory, but it would be necessary to be able to group these positions under more fundamental headings and to know what has to be assumed to make any of the structure stand. It is not necessary to know how to perform an autopsy on the abortus products of a trisomy-detected pregnancy, but one does have to know what the pathologist is looking for to confirm the diagnosis and why that is important for the parents who choose to abort. The result of this real interaction should be an ethics and a science that change with each other and produce a new field, providing contributions that cannot yet be anticipated—not an exegesis, not a political reform, but an interdiscipline.

This final sense of interdiscipline is the primary rationale for this book by experts involved in the field of bioethics. We have tried to structure opportunities for questioning and modification among experts who have familiarity with both science and morality and who hope to make the sort of advance that genetics made when it combined the Mendelian laws of inheritance and the concept of evolutionary change with the knowledge of biochemistry of the genes to arrive at that promising and rapidly expanding interdiscipline of molecular genetics.

The introduction, by Charles Curran, provides an overview of the field.

Curran, coming from a long tradition in ethics-morality, presents a modern interpretation of morality based on the natural law position. According to his position, God's preordained plan for the universe is known to us provisionally, not with intuitive certainty, because of our imperfections and lack of omniscience. We do the best we can in discovering this plan, have faith that it is there to be found, and use our best knowledge of this natural law to consider issues of concern in science and medicine. Curran identifies what he considers those issues to be and how best to address them. He identifies the ethical assumptions, the ways of framing questions, and suggests what he feels are the best approaches. The consideration of medical issues leads back to an examination of the ethical perspective and to suggestions on how best to interpret that perspective: "Interdisciplinary dialogue about science and technology affords theological ethics an opportunity to critically examine some of its own concepts." While, from the theologian's perspective, natural law is not altered, examination of its application to medical problems can result in new interpretations that grow from experience with such application.

The interaction can proceed in the opposite direction as well. Medicine can be seen, not as primarily a technical profession but as it is ordinarily practiced, a moral endeavor, since its concern is with the welfare of the person. Medical ethics problems are not framed in terms of the moral versus the scientific but rather in terms of the moral and the scientific. All clinical judgments are moral judgments intrinsically. Because medicine is a human intervention or raises the choice of intervening, and because it usually involves alternative ways of intervening, the ordinary issues of case management are usually issues of value choices. Medicine can be viewed then as applied ethics, and ethics can be approached in the same concentrated, empirical way that has led to the continuing increase in medical and scientific knowledge. Medicine is ethical behavior, but it is sometimes uninstructed ethical behavior. What medical ethics, relevantly developed, can do is to make that behavior informed and explicit. The several physicians who have contributed to this book demonstrate this third sense of interdiscipline.

It is the ability to modify or even radically change paradigms that characterizes science. There are times when the facts just will not fit, the explanation falls flat, and a reevaluation is in order. If bioethics is to be interdisciplinary in the sense we have suggested, it will need to somehow incorporate this methodology; and here is a major problem. Ethics has usually been viewed as conceptual or a priori and not empirical, as "ought" and not "is." One of the challenges of this sense of interdiscipline will be to show that values can be derived from facts and that facts are value laden.

For science and ethics to work together, then, we need to put aside two-handed thinking: on the one hand, science or medicine; on the other hand, ethics. The possibilities of a bioethics interdiscipline present some important, fundamental issues. These issues reappear throughout the chapters of this book, as specific challenges raised by scientific and medical advances are discussed.

Science and Morality

1 Science, Morality, and the Human Future

Charles E. Curran

The context of an interdisciplinary symposium on science, morality, and the human future explains the purpose and outline of this book. This chapter approaches the question from the perspective of Christian theological ethics in the Roman Catholic tradition. Since this chapter is the opening presentation, the first section describes some of the questions that are discussed later in greater detail. However, the primary purpose of the first section is to explain why these problems have come to the attention of our society. The second section, from the perspective of Christian ethics, proposes an approach for understanding and evaluating science and technology in relation to our human existence. The third section considers some important concepts in Christian ethics and how they shape an approach to the question of science, morality, and the human future. These concepts are not static but are very much influenced by experience and ongoing dialogue.[1]

The Problems: What and Why

This section surveys some of the problems and questions that are discussed in greater detail in the course of this book. Above all, this discussion focuses on an attempt to explain why these problems have arisen and become so acute today. The major contention maintains that these questions have arisen because of the tremendous advances that have occurred in science and technology. Science is understood in the broader sense of the empirical knowledge of the human; technology is defined as applied science by which human beings are able to control and influence human existence. Human beings have a greater power than ever before to control and influence their human existence because of the advances in science and technology.

Great progress has been made in medical and genetic knowledge and their technologies in our own lifetimes. Consider the phenomenal progress in the one area of drug therapy: The antibiotics, the antihistamines, and the psychoactive drugs, three of what are now the eight major classes of prescribed therapeutic drugs, were unknown forty years ago. The sulfas and the vitamins, two other major classes of drugs, were introduced between the two world wars. Barbiturates and hormones were discovered somewhat earlier in the century. Before this century only narcotic drugs were known but today's representatives of this class, with the exceptions of morphine and codeine, are recently developed drugs.[2] A momentary reflection calls to mind the startling advances in all aspects of medicine in our day—for example, the control of fertility, complicated heart transplants and operations, and a greatly increased life expectancy. However, these biomedical developments have also brought with them new and perplexing ethical dilemmas.

Since 1960, there has been an ever-growing interest in biomedical ethics. Before 1960, little or no attention was paid to medical ethics by most people and disciplines. Few ethical problems existed because the criteria of good medicine and of good ethics were one and the same. The whole purpose of medicine was to care for and cure the individual patient. Whatever the doctor did was directed to the goal of helping the individual patient. Ethics proposed exactly the same criterion—that is, any medical procedure, therapy, or treatment is good if it is to the benefit of the individual patient. Since both medicine and ethics recognized the same basic criterion, there were few, if any, areas of conflict.

The very fact that contemporary developments in medicine and science give us greater knowledge and power constitutes a general reason for more awareness of ethical problems about our use of this power. However, these newer developments in many areas raise an entirely new set of ethical questions precisely because the purpose and goal of biomedical knowledge, power, and technology are no longer restricted to the cure and care of the individual patient. The following questions, which are discussed in greater detail in later chapters, illustrate the basic contention that ethical problems have arisen because of the greater knowledge and power that human beings possess and because, through this power, biomedical interventions no longer are always destined to help the individual who is somehow exposed to danger or even harmed in the process itself. The purpose of this section is not to solve all these problems but rather to give intelligibility and understanding to the more basic question of why these problems constitute ethical questions and dilemmas.[3]

First, consider transplants between living human beings. One of the great medical advances in the decade of the 1950s was the successful transplant of a kidney from one living person to another living person who had no kidney function and who needed a kidney in order to live. The donor had two kidneys so that the loss of one, while exposing the individual to some possible future danger, did not constitute any immediate, grave danger for the donor. Because of the problem of rejection in transplants, the donations usually came from a twin or a sibling of the person who needed the kidney. Through such a donation, the kidney recipient was able to live.

Kidney transplants marked a tremendous achievement of medical science and technology. However, kidney transplants between living persons also raised an entirely new ethical dilemma. Perhaps for the first time in medical history, some harm was being done to a person by medical technology, not in order to help that particular person but in order to help another. Harm was done to X in order to help Y. No longer was the unique purpose of medical technology the good of the individual patient. The new ethical problem created by this advance was not insoluble, but it indicated the types of problems that would be increasingly faced in the future.[4]

Experimentation is a second illustration of the basic contention that biomedical progress has brought with it new ethical dilemmas. The phenomenal progress of contemporary medical science and technology would have been impossible without experimentation. It is very important to recognize the distinction between therapy and experimentation. Therapy is done for the good of the individual patient; experimentation is done not for the good of the individual subject but for the good of medical knowledge or some future patients. Sometimes a doctor might suggest to the patient the use of an experimental drug, but this is not experimentation in the strict sense of the term. In this case the whole purpose of taking the drug is to cure or to help the individual patient. The ethical problem in experimentation comes from the fact that the individual subject of the experimentation is exposed to some harm, not for the individual's good but for the good of science or some future patients. Without experimentation, progress in medical technology and drugs would be practically impossible, but abuses connected with experimentation have often been documented.[5] Recall the experimentation with people suffering from syphilis in Tuskegee, Alabama, in the late 1930s and the 1940s; even after penicillin was recognized as the ordinary treatment for syphilis, it was not given to those people who were involved in this experiment.[6]

The fundamental ethical concern in experimentation is to protect the freedom, dignity, and autonomy of the subject of experimentation. As a free agent, the subject must cooperate with the researcher in the quest of new knowledge. The subject is a free and autonomous person and not just an object. To protect the person and dignity of the subject, ethicists and lawyers call for the need for informed consent as a basic minimum. However, in practice, there are many obstacles to obtaining informed consent. Statistics reveal that the vast majority of experiments occur in the charity wards of hospitals and not in the private rooms. A person in the hospital, and especially a poor person, is very vulnerable. Many people would find it impossible to say no to what the doctor wants. To safeguard the rights of the individual person, then, it is very important to distinguish in theory and in practice between therapy and experimentation, between the doctor and the researcher. Not everyone wearing a white coat in a hospital is a doctor whose primary concern is the good of the individual patient. Experimentation can be morally acceptable under the proper conditions, but the rights of the subject can readily be infringed upon.

Death and dying constitute a third area of moral dilemmas today; and, again, medical progress has made the problems more acute.[7] Medical technology today can prolong the life for an indefinite period. In a previous generation there were no respirators, to say nothing of the whole phalanx of technologies that are used in the modern hospital. Is there a moral obligation to do everything possible to keep human life in existence? In our society, all of us in one way or another, either with regard to ourselves personally or with regard to a loved one, will probably face the decision of whether or not to pull the plug. From my ethical perspective, there is no moral obligation to do everything possible to prolong human existence.

Doctors, lawyers, and ethicists today are raising the question about the definition of death or the criterion for determining if death has occurred. Until recently, everyone accepted the common criterion of the lack of breathing as a sign of death. However, that criterion is not always helpful today precisely because modern medical technology can keep breathing going indefinitely. A new criterion is necessary, and most of these new approaches center on brain death.

A fourth problem area brought about by medical progress concerns behavior control.[8] Through drugs and electrical impulses, deviant human behavior can be somewhat controlled. Human beings today possess this awesome power. Some have suggested behavior-control techniques as a substitute not only for mental hospitals but also for prisons. Obviously, attempts to control and regulate the behavior of others raise a host of ethical problems.

A fifth problem area that is also discussed later in this book concerns genetic engineering and human reproduction. The so-called test-tube baby is now a reality. Conception can occur outside the womb in a petri dish (not a test tube), and the conceptus can then be implanted in the womb of a woman and brought to birth. In the future, even more startling ways of reproducing may become actualities. These actual and possible procedures raise significant ethical questions about the nature of human parenthood, the beginning of human life, the problem of mistakes and mishaps, and the power of human beings to control their destiny.

A sixth and final consideration concerns priorities in medicine. The priorities question owes much of its acuteness to developing technologies and takes many different forms. One aspect of the priorities question concerns who will receive lifesaving treatment when only a limited number of technologies are available. Some people claim that we should not play God and make such life-and-death decisions. However, we have no choice; one way or another these decisions will be made. My primary ethical concern is to safeguard the basic equality of all human beings in making these decisions. After medical considerations are taken into account, the basic equality of all human lives can ordinarily best be safeguarded by a lottery or a first-come-first-served method of selection. Otherwise, we become involved in the unacceptable role of deciding that some human lives are more valuable than others.

Another important aspect of the priorities question concerns the investment of our funds and our talents. Should we spend huge amounts on exotic lifesaving technologies for a few, or should we invest in more mundane realities such as maternal-health-care and infant-mortality problems? Should more concern and funding be directed to preventive health care rather than to forms of cure? As a society, do we not have an obligation to make sure that all our people have access to basic medical care? These are some of the important priority questions that we as a nation must face.

This first section has provided an overall view of some of the more important ethical questions raised by the technological progress that has occurred in science and modern medicine. In addition to raising the problems, the purpose was to give some understanding about why and how these problems have arisen, but no attempt has been made to solve or discuss them in any detail. Subsequent chapters deal with many of these topics.

Theological Ethics Looks at the Scientific and the Technological

From the perspective of theological ethics in the Catholic tradition, this section proposes a framework for understanding and evaluating science and technology, especially biomedical science and technology. Three theses are developed: (1) Theological ethics must be open to the data of science and the possibilities provided by technology; (2) too often in the past, practical suspicion characterized the reaction of the Catholic tradition to science; and (3) although important, the scientific and the technological are limited and not totally identified with the human. These three theses attempt to establish an approach to understanding the importance and the limitations of science and technology. Much debate occurs in the world today about whether science and technology are good or bad.[9] My own approach tries to avoid simplistic extremes. These three theses point to the fact that science and technology constitute human, but limited, goods.

The first thesis maintains that theological ethics must be open to the data of science and to the possibilities of technology. My understanding of Christian ethics (obviously Christian ethicists, like economists and other scholars, do not always agree among themselves) posits a basic openness to human reason, human science, and the possibilities for human betterment afforded by technology.

The first reason for a basic openness to science and technology stems from an acceptance of the goodness of human reason and the recognition that faith and reason can never contradict one another. Some religious and theological traditions are quite wary of human reason, but the Roman Catholic tradition has, at least in theory, insisted on the goodness of human reason and its ability to arrive at true wisdom and knowledge. The medieval theologians strongly insisted that faith and reason can never contradict one another. Faith is God's gift to us, but so is human reason.[10] Some of the first scientists were themselves theologians—for example, Albert the Great and Roger Bacon. Catholic tradition has always been willing to sponsor universities and the search for the truth through human reason. Thomas Aquinas, the most significant figure in the Catholic theological tradition, well illustrates the harmony that the tradition asserts as existing between faith and reason. The genius of Aquinas was to take the thought of a pagan philosopher who never knew Jesus and to use this philosophy to explain and understand better the mysteries of the Christian faith. Reason and science are not opposed to our Christian understanding, but rather they can enhance and contribute greatly to our understanding of the human.

A second reason for a theoretical openness to what science can discover and to what technology can do comes from a theological recognition of the goodness of the natural and of the human. Christian theology has constantly struggled with the problem of the relationship between the divine and the human, between grace and nature. The Catholic tradition has insisted that there is no incompatibility between grace and nature. Grace does not destroy nature but rather builds on nature. The older Catholic understanding of the relationship between grace and nature must be criticized for a number of reasons, but its basic acceptance of the goodness of the natural and of the human remains true.

There are theologies that, for various reasons, do not accept the basic goodness of the natural and of the human. Some theologians maintain that sin has utterly marred and destroyed the goodness of the human. In my judgment, as is explained later, some theological traditions, including the Roman Catholic, have not given enough importance to the reality of sin, but sin can never totally negate the goodness of the human that is created by God. Sin infects and affects but does not destroy the goodness of the human. Other theologians downplay the human and human capacities because they want to stress more the power, the glory, and the majesty of God. It is a false dilemma to think that, in order to emphasize the glory and the power of the divine, one has to minimize the glory and the power of the human. Precisely through the gift of creation human beings share in the power, the goodness, and the glory of God. To affirm the power and greatness of God one does not have to downplay the human. Participation and mediation aptly describe the relationship between the divine and the human. Thomas Aquinas based his whole ethic on the human being, who is an image of God, precisely on the basis of intellect, free will, and the power of self-determination.[11] My theological perspective affirms the goodness of the human and the goodness of reason with a consequent openness to the discoveries of science and to the possibilities provided by technology.

The second thesis is basically historical and recognizes that the Catholic theological tradition has not always lived up to its theoretical affirmation that there can be no contradiction between faith and reason. Too often in the midst of tensions, this tradition has lost the courage of its basic assertion that faith and reason cannot contradict one another. At times there grew up suspicion or even hostility between Catholic faith or theology, and science. Looking back in history, we find many examples of this suspicion and hostility toward science. Think, for example, of the Galileo case. In the nineteenth century, theology, and especially Catholic theology, felt threatened by the teaching of biology on evolution. Later there seemed to be a great opposition between theology and psychiatry and psychotherapy. Theologians in the twentieth century were often suspicious of the application of the tools of historical criticism to the Scriptures. There is no need to delay any longer over the thesis that historically there has often been suspicion, and at times even hostility, between theology and the human sciences. However, this attitude was in basic contradiction to the assertion that faith and reason cannot contradict one another.

The third thesis recognizes the limitations of science and technology. A perfect identity does not exist between the human and the scientific or technological. Science and technology must always be in the service of the truly human. A first limitation comes from the fact that the human embraces much more than the scientific and the technological. The human realities of freedom, wonder, awe, and suffering all transcend the empirical and the technological. Empirical science and technology can never solve the deeper human problems of meaning, of relationships, and of life and death itself. Empirical science and technology can contribute much to the meaning and enhancing of the human, but important aspects of the human transcend the empirical and the technological.

Since the human is not totally identical with the scientific, the human sometimes must say no to the scientific and to the technological. A good practical illustration of saying no to the technological in the name of the human is shutting off the respirator and allowing the person to die in peace. Actually, the problem of whether human beings must do everything possible to keep human life in existence goes back a long time. In the sixteenth century the question was raised about the obligation to undergo surgery to save one's life if the operation involved excruciating pain. Remember that anesthesia was first used successfully only in the middle of the nineteenth century. The Catholic theological tradition accepted the conclusion that excruciating pain usually excuses one from the obligation of undergoing surgery even if surgery is the only way to save one's life.[12]

A second source of limitation comes from the limitation of any one particular science or technology. Many different sciences contribute to our understanding of the human. The human moral judgment must take into consideration all these different aspects of the human—the psychological, the sociological, the hygienic, the eugenic, the medical, and so forth. In our complex human reality, nothing is perfect from every one of these perspectives. The human judgment in a sense is the ultimate judgment that relativizes all the other particular judgments of the individual empirical sciences. The perspective of any one empirical science can never be equated with the total human perspective.

Two illustrations show the limitations of any one science in relation to the other scientific perspectives and to the truly human perspective. The first example concerns the science of sociology, which is interested in assembling data about how systems and people act and respond. However, sometimes in order to find out these data it might be necessary to invade the privacy of individuals or to deceive them. At times society should say no to possible sociological investigations precisely because they violate important human rights. In the name of the truly human, the sociological perspective must be controlled and, at times, even limited. There is some knowledge we do not have the moral right to acquire because of the bad means that must be used in the acquisition of such knowledge.

A second illustration comes from a subject to be discussed later in this book—namely, genetic engineering and reproductive technologies. Scientists today talk about the future possibility of cloning human beings. Cloning is most easily understood as the copying of human beings. One can produce genetic twins of the original by transplanting the nucleus of the cell. It might be biologically and genetically possible in the future to reproduce one thousand Einsteins or multiple copies of other famous and successful people. Think of the tremendous advantages available to the human race. However, on second thought, such a development might cause many problems. Consider, for example, the psychological problems that might ensue. Even now we know the problems that exist for children of famous parents. Problems also exist for twins who achieve their identity vis-à-vis one another. How are a thousand Einsteins ever going to achieve their own individuality as human persons? The biological and the genetic are not the only perspectives that must be considered in the whole question of cloning. These illustrations point up the limitations of any one empirical science that must always be related to all the other scientific perspectives and to the fully human perspective.

A third source of limitation comes from the different opinions within any one particular empirical science or technology. This problem is acute for all laypeople who lack scientific expertise in a particular area or field. In psychology, should I follow Skinner or Erikson? In economics, is Friedman or Galbraith correct? One of the great ethical dilemmas facing society today is the use of nuclear power. Are nuclear reactors safe? Within most disciplines (including Christian ethics), there exists a great diversity of opinions. The differences within one science add to the complexity of many of the problems that face the modern world. However, from our perspective, these differences of opinion show the limitations existing within any one science in terms of its relationship to the fully human.

This section has proposed a framework for understanding and evaluating the scientific and the technological. In brief, empirical science and technology might be described as limited goods. They are goods that can contribute to our understanding and developing of the human, but they are also limited and must constantly be under the control and direction of the human.

Theological Ethics Looks at Itself

This section examines four important concepts in theological ethics that provide a framework for the approach to be taken to the questions involving science, morality, and the human future. Theological ethics is not unaffected by human experience, for the discipline is not an a priori construct. Interdisciplinary dialogue about science and technology affords theological ethics an opportunity to critically examine some of its own concepts. On the one hand, we are constantly trying to understand better our basic religious and theological concepts in the light of our experience. On the other hand, the danger always exists that human experience might distort some of these basic religious and theological ideas and principles. This section discusses four important concepts in theological ethics—(1) the understanding of God, (2) anthropology, (3) human progress, and (4) the proper ethical model. These important concepts have a great influence on how one approaches the question of science, morality, and the human future.

Our understandings of God and of how God works in the world are very much influenced by human experience. Ask yourself this question: Do you think of God as a Chinese woman? God can just as easily be a Chinese woman as a white, European, or American male. The Scriptures remind us that no one has ever seen God. Our understanding of God will always be colored by our own experience. However, at the same time we must always be careful not to distort the concept of God because of our own limited experience. God may surprise us all when she finally reveals herself! There is always the danger of committing the great idolatry—that is, making God into our own image and likeness. The Christian United States has too often identified God with its own national goals, aims, and adventures.

In terms of theological ethics, an older concept tended to see God and God's role in the world in a static way. In both the Catholic natural law approach and the Lutheran theory, God was seen primarily in terms of ordering our world. From this it was easy for theological ethics to talk about the eternal and immutable plan of God for our world. One of the dangers with such an approach is that what is a historically conditioned reality too easily becomes identified with the immutable plan of God. An older Catholic understanding called for the union of church and state as the ideal form of relationship. In reality, this abstract ideal was greatly influenced by the historical reality introduced by Constantine.

Contemporary theological ethics usually proposes a more active understanding of God's working in the world. Liberation theology, which was developed in a Catholic context in South America but which has also evolved in feminist and black perspectives, tends to see God acting in the world and in history to bring about the liberation and freedom of human beings.[13] The good news of the Gospel is the story of liberation from sin, but freedom from sin includes freedom from all the oppressive social, political, and economic circumstances that surround people today.

I am in basic agreement with the thrust of the image and role of God proposed in liberation theology, but I want to add some amendments or modifications. Liberation involves a long, hard struggle, and it will never be fully accomplished in this world. The conclusion from this modification is not that Christians should accept the status quo but that Christians must recognize that the struggle for liberation will never be totally successful in this imperfect world in which we live. Freedom is a multifaceted reality and embraces many different aspects. At times it is easy to forget certain aspects of unfreedom, especially in our own lives. In our own society, the teenagers are usually held up as the freest and most liberated people in society. If this is true, why do they have a dress code that is stricter than that of any private boarding school of the past generation? Why do all teenagers think they have to wear jeans? Latin American liberation theologians see oppression primarily in economic realities; feminist theologians often see oppression primarily as sexist. There is a danger of limiting oppression to just one area and forgetting the other important areas in which it exists.

In general, a contemporary concept of God's role in the world stresses the humanizing and freeing activity of God to bring about peace and justice. Such an approach is usually acceptable, but one must always be alert to the danger of too easily identifying God's own work with our own narrow perspectives or particular causes. In addition, the recognition of a dynamic presence of God in history cannot deny the important continuities that are a part of human existence. The danger of forgetting that the fullness of God's work and Kingdom will only come at the end of time has already been pointed out. Later considerations will call attention to the continuing presence of sin in our world.

Anthropology is another very significant and fundamental concept in Christian ethics. One important aspect of this question involves either an optimistic or pessimistic view of the human. The diversity in theological ethics includes both extremes. Harvey Cox, in the 1960s, proposed a very optimistic anthropology. Cox does not forget the reality of sin, but for him the primary human sin is *acedia*, or sloth—the failure to take responsibility for the world in which we live. The great temptation of human beings is "to leave it to the snake" and not take initiative and responsibility for the world. Sloth underestimates human abilities and powers.[14] Conversely, Paul Ramsey, a Methodist theologian from Princeton, maintains that the great sin of human beings is pride. Human beings tend to claim too much for themselves and to forget about their inherent limitations and sinfulness. Pride, or claiming too much power, is the root of all the evils in the world. The great human temptation is to play God.[15]

My approach to anthropology is based on the stance that I propose for Christian ethics. The Christian looks at the world and at all reality in the light of the fivefold Christian mysteries of creation, sin, incarnation, redemption, and resurrection destiny. Such a stance recognizes a basic goodness about human beings grounded in the mysteries of creation, incarnation, and redemption. At the same time, the presence of sin infects and affects all of human reality, and the future of resurrection destiny reminds human beings of our incompleteness and limitations in this world. Such a stance, or perspective, avoids the extremes of optimism and pessimism and grounds what might be called a critically open perspective. Power and technology can be used for good, but the dangers of misuse are always present. Such a realistic stance reminds contemporary human beings that genetic planners are no more immune from the finitude and sinfulness of human life than public-policy and foreign-policy planners.

From a more philosophical perspective of anthropology, some contemporaries see the human person as a self-creator.[16] Such a concept recognizes that, through science and technology, human beings have acquired great power over human existence. No one can deny that human beings have greater power than ever before precisely because of many technological advances, but the image of the human person as a self-creator goes too far. The creator is the one who makes something out of nothing. Unfortunately, but in my judgment it is more accurate to say fortunately, human beings start with something and not with nothing. We start with the realities of our embodied selves and the world in which we live. Human beings exist among many limitations. We live in only one space in the universe and in one comparatively short period of time. We tire and become fatigued. There are many things we cannot do because of our limitations. Yes, human beings are creative, but they are also limited. Consider for a moment our ability to speak and communicate. This ability to speak signals the transcendence and the creativity of human beings, but still our ability to communicate through speech is quite limited. Human beings must use many words, and all of us have experienced the frustration of not being able to communicate how we really feel or believe. Human beings are both creative and limited.

An anthropology that understands the human person in the image of a self-creator tends to see human activity only in terms of freedom and to reduce all ethical issues to freedom. Freedom remains a very important ethical consideration, but it is not the only one. The fact that the person involved or all persons concerned agree to something does not make it right. There are other ethical realities in addition to our freedom. The stress on freedom alone also fails to recognize the many limitations that characterize human existence. The ecological crisis and the population problem remind human beings of limitations. Human beings are not free to intervene any way they want in the environment. Complex ecological systems exist that must be respected. There are limits to what we can and should do. A contemporary emphasis on small is beautiful has arisen as a reaction to the failure to recognize the limits to creativity and freedom.

A third important concept in Christian ethics is the understanding of human progress. No one can deny the tremendous technological progress that has occurred. The progress in biomedical science and technology was mentioned at the beginning of this chapter. Consider the fabulous technological progress in transportation. In one generation, human beings went from the horse-drawn carriage to the space flight. At first glance, technological progress seems to be always progressive. The newer development builds upon and improves on the older. Such progress appears to be constantly going forward, each new discovery becoming another step on the way to utopia.

At the very minimum, technological progress is not the same as truly human progress. Human progress does not occur in such a constantly developing, ever-improving direction. Why is it that we still read Shakespeare, listen to Beethoven, and admire the paintings of Giotto or van Gogh? Too often the allure of technological progress deceives human beings into thinking that truly human progress involves a constant improvement, the newer always better than the old.

A deeper reflection recognizes that technological progress itself is ambivalent and does not involve progressive growth. In the enchantment with technological progress, many human beings forget about the negative effects of that progress. The pollution of our air, the impurity of our waters, the screeching sounds of cars and airplanes in our ears, and the devastation of our ecological resources all remind us of the ambivalence of technological progress.

There is an even more fundamental question to raise about human progress: Is there such a thing as truly human progress in our world? A glance back over the history of human existence seems to give contradictory evidence. In more primitive times human beings threw stones at one another; in the modern world human beings are prepared to hurl deadly nuclear weapons at fellow dwellers on planet earth with the possible consequence of even extinguishing human life in apocalyptic destruction. Conversely, there seems to have been some progress in our recognition of the freedom and dignity of human beings with a basic set of inalienable rights. Both from historical observations and from my theological stance with its fivefold Christian mysteries, I think there has been some truly human progress in this world. However, this progress is slow, limited, difficult, and often marred with regressions along the way. The Christian gospel calls human beings to strive for true progress and a greater presence of the Kingdom in this world, but the fullness of the Kingdom will only come at the end of time.

A fourth important concept in Christian ethics concerns the model employed for understanding the Christian life. It is impossible to develop this question at great depth here. The only purpose of this consideration is to show that a technological model is inadequate for understanding Christian and human ethical life. The technological model of human existence sees the person primarily as a maker or a doer. The technological model emphasizes the control and power of the human being who is a maker. The human being has the ability to shape and direct all that comes into his hands. Take the simple example of the carpenter shaping wood into a table. However, both theology and human experience remind us that we do not have such control over our human lives and existence. Many things happen to us in our life that we do not want. Human existence would probably be quite dull if we had full power and control over our lives. There is a great difference between a technological model of human existence and a proper theological, ethical understanding.

The technological model cannot come to grips with the reality of human suffering. Again, human experience and theology both recognize the problem of suffering. The book of Genesis was originally written not to explain the reality of creation but to come to grips with the problem of how evil and suffering can exist in the world in the light of a belief in a good and gracious God who made all things. Evil and suffering raise a continual problem for believers—for Job, for Jesus, for the survivors and witnesses of Auschwitz. The technological model of human existence, with its emphasis on human power and control, cannot deal with the reality of suffering. The Christian perspective sees suffering in the light of the Paschal Mystery of Jesus. Jesus gave a redemptive quality to suffering. Through the suffering of the Cross, Jesus overcame the power of evil and sin.

The Christian tradition recognizes the reality of suffering, sees suffering in the light of the Cross and resurrection of Jesus, but ultimately must admit the mystery of suffering. Likewise, the Christian tradition must honestly admit that its own approach to suffering in the past has at times been deficient. The Marxist critique of religion and of Christianity contains some truth. All too often Christians were given the advice to put up with their suffering and to endure the present vale of tears because their reward would be great in heaven. Faced with the reality of evil, Christians are called upon to struggle to overcome it, but in this world, evil and suffering will always continue to exist. A technological model has no place for the reality of suffering. A Christian vision tries to deal with the reality but in the past has occasionally proposed distorted and one-sided approaches.

The technological model of human existence is interested in results, effects, and success. When applied to human existence, such a model sees human value in terms of what one does, makes, or accomplishes. The whole of the Judaeo-Christian tradition cries out against such an understanding of human value and dignity. Who are the privileged people in the Kingdom of God? Are they the rich, the powerful, the doers? No, the privileged people in the Kingdom of God are the poor, the lowly, the outcast, and the children. Human values and dignity do not ultimately depend on what one does, makes, or accomplishes.

A technological model applied to human existence tends to result in a total ethical consequentialism. The best illustration of such a strict consequentialism is utilitarianism with its criterion of the greatest good of the greatest number; but too often the greatest good of the greatest number leaves out the weak, the handicapped, the poor, and the oppressed. In my judgment, a Christian approach to human existence and ethics cannot accept such a utilitarian approach.

This entire consideration of the differences between a technological model applied to human existence and a Christian approach well exemplifies what was said earlier about the technological. Science and technology constitute limited goods, but the human and the Christian involve more than just the scientific and the technological.

Notes

1. Many aspects touched upon in this chapter are developed in my book *Issues in Sexual and Medical Ethics* (Notre Dame, In.: University of Notre Dame Press, 1978).

2. Bernard Barber et al., *Research on Human Subjects* (New York: Russell Sage Foundation, 1973), p. 1.

3. Two important reference tools in the area of bioethics should be mentioned: Warren T. Reich, ed., *Encyclopedia of Bioethics,* 4 vols. (New York: Free Press, 1978); and LeRoy Walters, ed., *Bibliography of Bioethics* (Detroit: Gale Research Co., 1975).

4. James B. Nelson, *Human Medicine: Ethical Perspectives on Medical Issues* (Minneapolis, Minn.: Augsburg Publishing House, 1973), pp. 149–170.

5. For an exhaustive study of human experimentation, see Jay Katz, *Experimentation with Human Beings* (New York: Russell Sage Foundation, 1972).

6. R.H. Kampmeier, ''Final Report on the 'Tuskegee Syphilis Study',''' *Southern Medical Journal* 67 (November 1974):1349–1353.

7. See, for example, Robert M. Veatch, *Death, Dying and the Biological Revolution: Our Last Quest for Responsibility* (New Haven: Yale University Press, 1976).

8. For a discussion of behavior control from various perspectives, see Willard Gaylin; Joel Meister; and Robert Neville, eds., *Operating on the Mind* (New York: Basic Books, 1975).

9. In 1979, the World Council of Churches sponsored a Conference on Faith, Science and the Future, which was held at the Massachusetts Institute of Technology, 12–24 July. This conference and the publications connected with it discussed in great detail and from many different perspectives the understanding of science and technology in the light of Christian faith. For the results of the conference, see Roger L. Shinn, ed., *Faith and Science in an Unjust World,* vol. 1: *Plenary Presentations* (Philadelphia: Fortress Press, 1980); Paul Abrecht, ed., *Faith and Science in an Unjust World,* vol.2: *Reports and Recommendations* (Philadelphia: Fortress Press, 1980); and Abrecht, ed., *Faith, Science and the Future* (Philadelphia: Fortress Press, 1979), which is a preparatory volume published before the conference.

10. Vatican Council I, 3rd sess., Dogmatic Constitution, *Dei Filius,* Cap. 4 *De fide at ratione,* in *Enchiridion symbolorum definitionum et declarationum de rebus fidei at morum* 32nd ed., edited by H. Denzinger and A. Schömetzer, note 3017 (Barcelona: Herder, 1963). See also Fifth Lateran Council, Denzinger, note 1441.

11. Thomas Aquinas, ''Prologue,'' *Summa Theologiae,* I^{a} II^{ae}.

12. Daniel A. Cronin, *The Moral Law in Regard to the Ordinary and Extraordinary Means of Conserving Life* (Rome: Pontifical Gregorian University, 1958).

13. The best known examples of liberation theology in the South American context are Gustavo Gutierrez, *A Theology of Liberation* (Mary-

knoll, N.Y.: Orbis Books, 1973); and Juan Luis Segundo, *The Liberation of Theology* (Maryknoll, N.Y.: Orbis Books, 1976).

14. Harvey G. Cox, *On Not Leaving It to the Snake* (New York: Macmillan, 1967), especially pp. ix–xix.

15. Paul Ramsey, *Fabricated Man: The Ethics of Genetic Control* (New Haven: Yale University Press, 1970), pp. 90, 151–159.

16. For a discussion of the creativity and freedom of human beings in the light of modern genetics, see John J. McNeill, "Freedom and the Future," *Theological Studies* 33 (1972):503–530.

Part I
Interventions at the Beginning of Life: Reproductive Choices and Genetic Knowledge

Some of the most significant scientific advances in the last three decades have been in the fields of molecular genetics and cell biology. Among other things, this new knowledge has provided insight into the human genetic endowment and the very earliest developmental events in the human life cycle. It has also made possible significant manipulations of the genetic material. The term *genetic engineering* has been invoked to describe applications or potential applications of this work, such as selecting for or against naturally occurring genes or groups of genes (that is, positive and negative eugenics) or endowing organisms with new characteristics.

Many of the current concerns, now part of the bioethics agenda, have been generated by the increased amount of information that can be obtained about human genes. It is now possible, with the use of biochemical and genetic techniques, to determine the presence or absence of specific malfunctioning human genes. Thus, individuals who are carriers of genes for some serious disorders can be identified through the use of a variety of testing procedures. These individuals, though healthy themselves, have a 50 percent chance of passing the gene on, at each pregnancy, to their offspring, who will then also be carriers of the harmful gene. If both partners should carry the same abnormal gene, they have a 25 percent chance of having offspring with the disorder at each pregnancy. Though carrier (or heterozygote) detection is becoming possible for an increasing number of inherited single-gene disorders, mass screening has only been conducted in two populations in which a greater risk exists for inheriting a specific genetic disorder—namely, in Ashkenazic Jewish communities for Tay-Sachs carriers and in black communities for carriers of the sickle hemoglobin gene that, when inherited from both parents, results in sickle cell anemia. Otherwise, carrier screening is currently employed only within families whose genetic history indicates that members may have a high risk of being carriers of genes for specific rare, heritable disorders. For the majority of inherited conditions caused by recessive genes, no carrier-testing procedures exist.

Genetic information can also be obtained at birth to identify those newborns who have inherited genetic illnesses for which effective treatment is possible, especially when physiological damage can be markedly reduced or avoided by prompt treatment. For example, phenylketonuria (PKU) disease—in which metabolism of the amino acid phenylalanine is impaired, leading to severe irreversible mental retardation—can be successfully treated by altering the diet to reduce intake of phenylalanine. In some states, as many as fourteen different disorders are being screened for at birth.

Perhaps the most dramatic advance in human genetics has been the development of techniques that can provide genetic information well before mid-pregnancy. The major method used to obtain such information is that of transabdominal amniocentesis performed early during the second trimester of pregnancy, usually at about the sixteenth week of gestation. Amniocentesis is the procedure in which a small amount of fluid (about 20 ml, two-thirds of an ounce) that surrounds the fetus is removed with a fine needle inserted into the amniotic sac through the abdominal wall (see figure 5–3). This fluid contains chemical substances produced by the fetus as well as cells that have been shed from the surfaces of the fetus. Both the cellular and fluid components can provide important genetic information. Though most of the fetal cells are nonviable, a very few (about one cell of every 2,000) are still alive and can grow under special conditions in the laboratory to produce a population of identical cells that can be studied. With appropriate methods, the chromosomes in these cells can be visualized, photographed, and organized into a pattern called a karyotype (figure I–1).

The karyotype allows the counting of the total chromosome number in each cell, identification and examination of the structure of each individual chromosome, as well as determination of the sex of the fetus. Deviations from the normal chromosome complement (23 pairs, 46 total chromosomes) or alterations in the amount of chromosomal material can cause severe developmental problems. The presence of one extra of the number 21 chromosome (see figure I–2) yields a condition known as trisomy 21, or Down's syndrome, characterized by mental retardation and some physical abnormalities. Other rarer syndromes such as Turner's, Edwards', and Klinefelter's are also caused by deviations from the normal 46. The incorrect total number most often results from occasional errors in gamete formation so that sperm or egg cells are produced with more or fewer than the correct complete set of 23 chromosomes. There also is a positive correlation between advanced maternal age and the occurrence of such chromosomal abnormalities (see figure 5–1). Even when the correct number of chromosomes is present, the incorrect amount of genetic material on a chromosome can also cause serious problems. The absence of a small portion of chromosome 5, for instance, results in *cri-du-chat* syndrome, in which affected individuals have profound mental retardation, a distinctive catlike crying sound, and facial abnormalities.

Sex determination, apparent from the karyotype by examination of the sex-chromosome pair, provides important information for predicting occurrence of genetic disorders brought on by genes located on the X chromo-

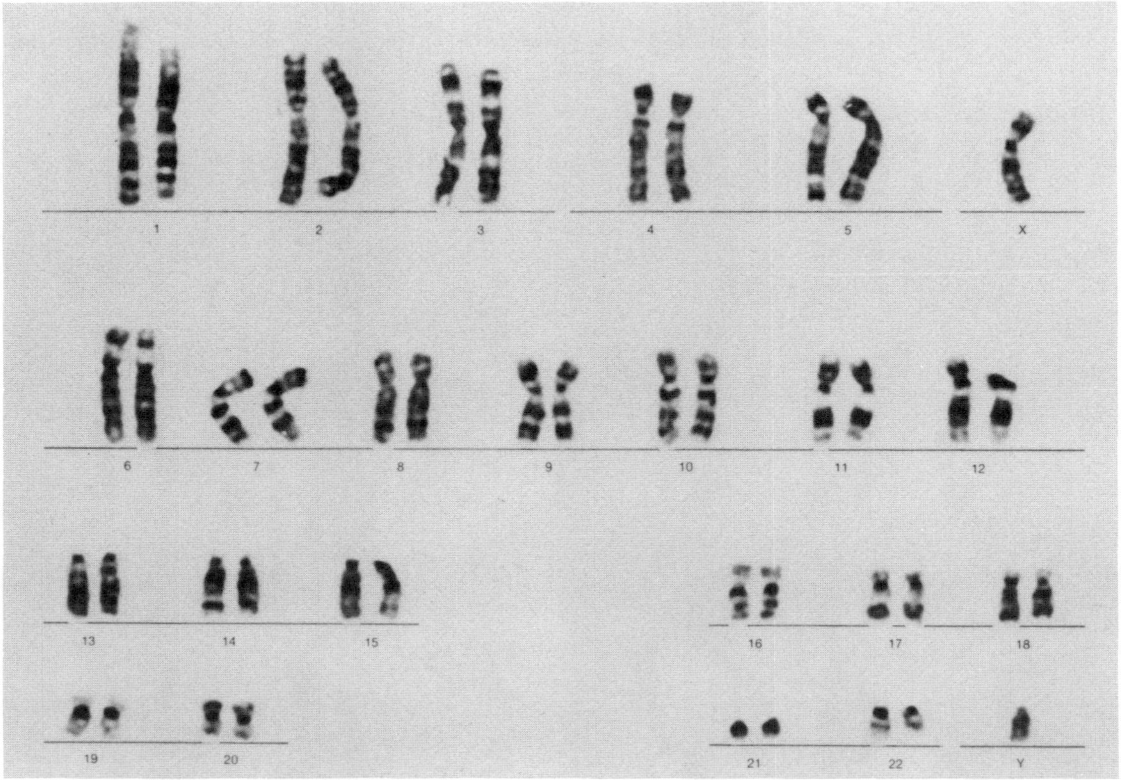

Note: The presence of an X chromosome and a Y chromosome indicates that the sex is male. The staining technique used results in the appearance of bands on the chromosomes. The banding pattern allows members of a like pair to be distinguished from other chromosomes of the same size.

Figure I–1. Karyotype Showing the Normal Chromosome Complement of 46 Chromosomes (23 Pairs)

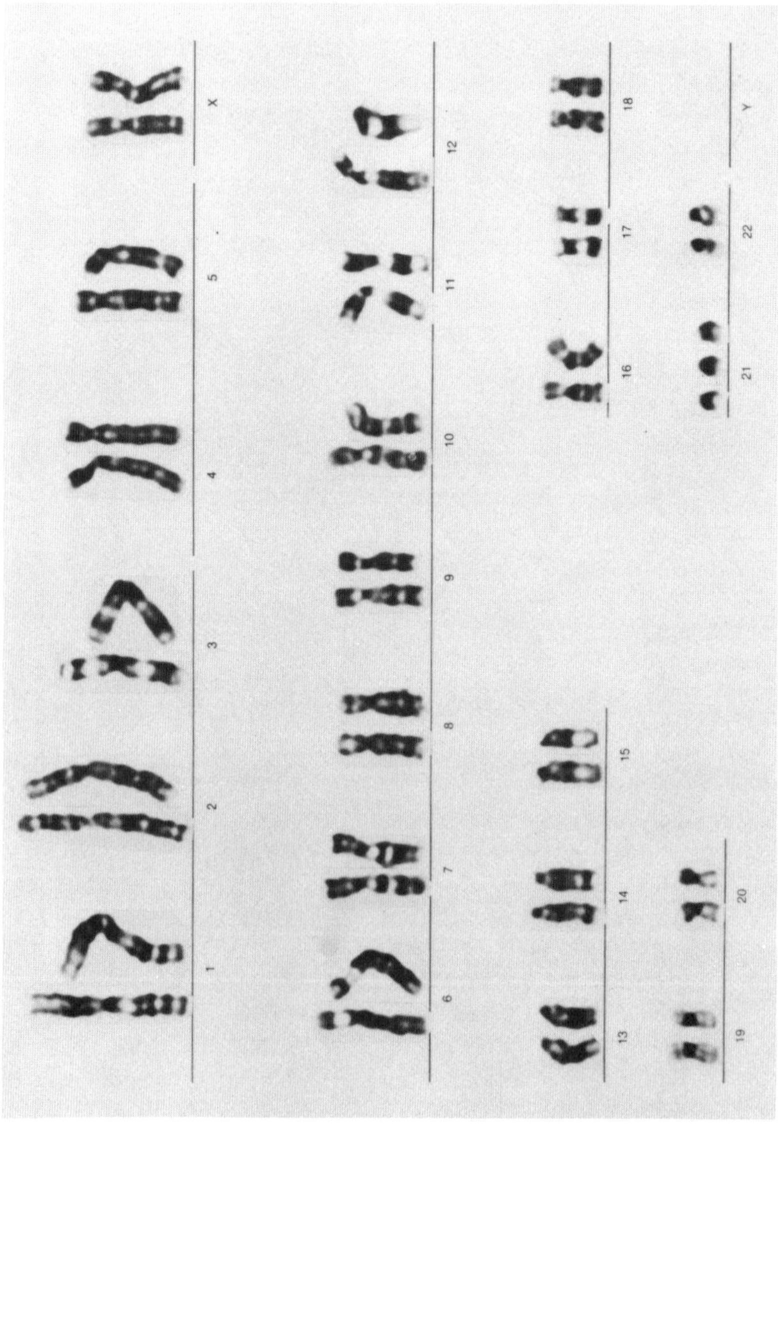

Note: The presence of two X chromosomes indicates that the sex is female. The total chromosome number here is 47.

Figure I–2. Trisomy 21 (Down's Syndrome) Karyotype Showing an Extra No. 21 Chromosome

some. In this case, males (who possess only one X chromosome) have a higher risk for inheriting the disorder. On average, one-half of the male offspring of a carrier mother will inherit the X chromosome bearing the defective gene. Female offspring of such a carrier are at no risk for developing the disorder, though each has a 50 percent chance of inheriting the gene and being carriers like their mother. (An illustration of sex-linked inheritance patterns is given in figure 5–2).

In addition to providing information about the chromosomal makeup, the fetal cells can be used to measure specific enzymes or cellular substances. The quantity of such substances is often an important clue to the presence of a gene disorder. For example, a lack of hexosaminidase-A enzyme in amniotic fluid cells establishes that the fetus has inherited Tay-Sachs disease. More than 75 other genetic disorders, including Hurler's syndrome, galactosemia, cystinosis, and Lesch-Nyhan syndrome, can currently be detected using fetal cells obtained at amniocentesis.

The cell-free fluid portion is also a useful source of chemical information about the fetus. An elevated level of the fetal protein, alpha-fetoprotein, has been correlated with spina bifida and other neural-tube defects. Measurements of other components in the amniotic fluid may provide information about myotonic dystrophy and adrenogenital syndrome.

Increased understanding of amniotic fluid chemical composition and amniotic fluid cell gene expression will permit other fetal characteristics to be determined. New technologies such as fetoscopy and ultrasonography are coming into wider use and will enlarge the potential to obtain more detailed information on the fetus. Still other technologies such as flow cytometry may allow information to be obtained much earlier than the sixteenth week of pregnancy. Prenatal intervention using "gene therapy" to replace missing or abnormal gene products or to surgically repair malformations has just started but in the future may become more common medical practice. However, those features or traits produced by the complex interaction of many different genes or those that develop as a result of environmental influences (intelligence, personality, musical ability, and so on) remain impossible to predict or alter.

Knowledge of the way the genome is organized and how it functions has been augmented recently by the increased ability to manipulate the genes themselves using recombinant DNA technology. DNA, deoxyribonucleic acid, is the chemical substance of which the genes are composed. With this new technology, small portions of DNA from one source can be recombined with, or inserted into, DNA from another source (usually bacterial). This provides an opportunity to study limited regions of the genetic material in order to acquire more insight into how the gene or genes so isolated function and how such function can be regulated. It also makes possible the assemblage in one organism of genetic capabilities from other organisms producing, for example, plants that might express bacterial genes, allowing them to fix nitrogen. Recombinant DNA technology has already had practical applications. *Escherichia coli* bacteria have been developed that have the capacity to produce needed human biological substances such as insulin and interferon. These techniques are also providing a new means for locating the exact position of genes on human chromosomes and for identifying specific forms of the genes present in the fetus. Extension of this technology to permanently alter or repair the human genetic material itself may be a longer-term outcome of this field.

Barton Childs, in chapter 2, makes a distinction between human and medical genetics that might be quite useful to keep in mind. While human genetics strives to enlarge our understanding, "to learn something about biological questions" (population genetics, chromosome mapping, developmental biology, recombinant DNA), medical genetics is concerned with health, with diagnosis, treatment, and prevention. The rapidly accumulating knowledge from human genetics directly impacts on medical genetics, and the difficulties that can arise when that knowledge is applied have raised questions about its value. Childs talks about this in terms of cost-benefit considerations, a method of analysis that has been applied in a number of areas such as environmental ethics. He identifies concerns about the specificity and sensitivity of tests, the network of who should be told, the genetic quality of the disease or its treatment, carrier testing, and confidentiality. He does a cost-benefit analysis for carrier testing and for antenatal (prenatal) diagnosis and concludes by emphasizing the value of broadening options, stressing that the choice of conditions under which techniques and knowledge are applied can maximize benefits and minimize costs and that human variability means the cost-benefit equation will be "weighted differently for different people." (How to handle uncertainty will be an ongoing problem.)

Ruth Hubbard approaches genetic decision making from a psychosocial and historical frame and uses a political perspective to develop her argument in chapter 3. She discusses the history of the eugenics movement, its shaky scientific foundations, and emphasizes how difficult it would be in any case to decrease the incidence of a harmful recessive gene in the population since most of these recessive genes are borne by healthy carriers. She also stresses the small risk we all share in procreating. In addition to her unease that medical genetics might take a social-agent role, she also has reservations about genetic counseling and amniocentesis at the physician/patient level, based on this perceived common risk and the goal of a perfect child. She feels such intervention could exaggerate our concern about genetic risks, unnecessarily increase our dependency on experts, and distract us from coming to grips with more pressing environmental risks.

Edmond Murphy specifically addresses the eugenic consequences of genetic engineering from the standpoint of the clinical and mathematical geneticist in chapter 4. In order to advance quantitative genetics and evolutionary theory, he points out that we need more than the tautologies he critiques in the first part of his chapter (''the survival of the survivors''). He tends to support Childs's distinction by indicating that eugenics can center on a ''strategic gain for the whole community'' (improving genetic stock) or a ''tactical gain for the individual family'' (having only healthy children). Murphy also maintains that evolutionary and genetic systems are very conservative—that is, they change slowly, and thus complexity or variance from a norm might have considerable adaptive value (or fitness). If genetic mechanisms, by their nature, are slow to respond to large environmental changes, it is advantageous to have a pool of many varieties of features in the genome (even if it appears to be ''junk'' for the present). While Murphy cautions that there may be realistic limits to the junk we can accumulate (''evolutionary survival demands good averages''), there is also a balanced need for a reservoir of diversity. For medical genetics this means that the very slow increase in the frequency of harmful recessive genes in the gene pool, through carrier screening or prenatal detection, is not an emergency problem. Dominant lethal genes are another story. Although the frequency of the gene in the population would be increased, if full fitness can be restored easily, the only problem would be if a social disaster occurred, making the direct treatment impossible. Murphy is more skeptical of positive eugenics. Finally, he feels genetic engineering will have little impact on this whole issue in terms of solving the problems that are evolutionary, biological, and ethical. On the grand evolutionary scale, he maintains that none of this will make much difference.

In chapter 5, John Fletcher begins by differentiating between morality and ethics. Morality refers to rules or systems that offer guidance when a specific conflict or dilemma exists. These rules are incumbent on everyone, and he disagrees with Joseph Fletcher's situational approach that regards them as only rules of thumb. Ethics refers to basic principles on which rules are made. At present, the two most widely used ethical positions are that of deontology, in which the emphasis is on one's duty to perform actions or deeds (truth telling, respect for personhood, the autonomy principle), and consequentialism, in which the goal is to act in a way that yields the best results (elimination of pain, increase in happiness). John Fletcher's major premise is that moral problems (which he carefully enumerates) posed by genetic screening, genetic counseling, prenatal detection, and abortion are best dealt with by the principles of freedom and fairness. He points out the contribution made by the Hastings Center (an institute for the study of science and values) in developing guidelines based on these principles. These principles incorporate elements of the deontological position (no one is forced to acquire genetic information about themselves or their offspring) and consequentialism (the methods involved are employed to relieve pain and suffering). He feels that this is the best available public policy because it allows the "coexistence of rival ethical views with the least amount of public punishment or repression," similar to religious toleration. This theme also runs through the previous chapters. Fletcher cautions, however, that such an ethical position is possible only in a stable and economically secure society.

The new developments in the field of human genetics have been rivaled by the increased ability to manipulate the germ cells—sperm and ova—that fuse to produce the zygote, or first cell of the new organism. Procedures for freezing sperm cells have been perfected, and artificial insemination now accounts for more than 10,000 births per year in the United States. In vitro fertilization, which involves manipulation of the ovum, has been developed to overcome infertility in those women whose fallopian tubes, the normal site of sperm-egg fusion, are blocked or absent. Pioneered by Patrick Steptoe and Robert Edwards in England, this procedure has, since 1978, been successful in a number of cases. Several children (known erroneously in the media as "test tube babies") have already been born, and a number of pregnancies are in progress. In chapter 6, John Biggers provides an in-depth review of this technique. In vitro fertilization is a multistep process in which a mature ovum, surgically removed from the ovary, is placed in contact with sperm cells in a laboratory dish, allowing fertilization to occur. If fertilization is successfully accomplished, a zygote is formed that proceeds by repeated divisions into the early embryonic stages of development. The early embryo is then returned to the uterus where attachment to the uterine lining and subsequent development can take place.

Indications from other animal systems are that hormonally induced ovulation may cause errors in chromosome number in the mature ovum and that high numbers of sperm cells surrounding the ovum in the laboratory dish, and the absence of the natural filtering out of defective sperm, could cause errors in chromosome complement in the zygote. No reliable data are available yet to evaluate whether the manipulations involved in this technology carry with them the risk of an increase in congenital defects in humans. According to Biggers, it would be expected that most abnormal embryos would be lost—as they are at significant levels during any normally produced pregnancies when chromosomal abnormalities, defective genes, or harmful intrauterine environmental factors are present—so that the overall risk may not be high.

Any intervention in reproductive processes poses some moral problems. Additional questions are raised by attempts to improve the low success rate of this procedure through hormonal treatment to increase the number of mature ova obtainable at one time (superovulation), freezing of several embryos to reduce the need for repeated surgical procedures, and use of extra embryos in fetal research. These questions have become public-policy concerns.

Warren Reich, in chapter 7, examines in vitro fertilization and embryo transfer in terms of public policy and ethics. He enumerates three milestone events that contributed to the formation of such public policy: (1) the birth of Louise Brown following in vitro fertilization; (2) the Del Zio legal suit that developed out of the deliberate destruction of Mrs. Del Zio's fertilized ovum prior to implantation; and (3) the *Roe* v. *Wade* abortion decision of the U.S. Supreme Court that declared that the fetus is not a person. All three share an important ethical question: "What are the moral and legal claims of the early, prenatal stages of human life?" These events precipitated a change in governmental regulatory and review policy that had amounted to a four-year moratorium on publicly supported research in this field. At the moment, no further policy decisions have been made. The governmental agencies have been reorganized, and the required advisory board no longer exists.

Reich outlines a variety of public-policy options: (1) direct restriction of in vitro fertilization, (2) refusal of government funding, or (3) imposition of procedural requirements such as an ethics review board. All of these options depend partly on "presuppositions about the relation of moral principles to public policy." How are such values and principles selected for policy decisions? What policy is good for society? Moral pluralism makes it difficult to answer these questions. Public-policy ethics tries to determine which among all the moral values should be protected. Reich then lists five criteria for the selection of values to be protected by public policy:

1. Policy should support what is arguably good and restrict what is arguably evil when required by the public order;
2. When policymakers must choose between two evils, they should choose the lesser;
3. Rights should not be restricted unless it is truly necessary for the public good;
4. Policy should not be excessively burdensome to fulfill or enforce;
5. In an impasse, a procedural resolution may be the only workable one.

Ethicists can assist by articulating the nature of the moral conflict and the principles, clarifying options in light of these principles and the public ethos, ranking moral options, and critiquing the trend of policymaking.

While Reich feels that some of the consequences of in vitro fertilization are wrong in terms of normative ethics, he sees no reason to ban its laboratory or clinical use. He discusses the meaning of parenthood, both conceptually and morally. He also reaffirms the union of procreation and marriage, feeling that responsible parenthood requires this, but he disagrees that procreation should be limited within marriage only to marital intercourse. In vitro fertilization, also, does not represent a separation of procreation from the sphere of love. The moral status of the human embryo is important too. Though the early embryonic stages command respect, Reich argues that they do not fully constitute a human being. The embryo should not be directly harmed or produced solely to be extinguished. However, clinical research that does not intend the destruction of embryos is consistent with respect for human life. He gives a number of conditions for the conduct of both therapeutic and nontherapeutic clinical research (within a fourteen-day moral option) and concludes with a discussion of risk of defect to the embryo and consideration of just allocation of resources, arguing that funding should be private, not public.

John Ladd, in chapter 8, considers that genetical ethics includes eugenics, gene therapy, genetic counseling, and genetic screening. Philosophically, he finds something of a mess: confused and confusing themes, myths and fairy tales, and pomposity. He identifies two overall attitudes toward genetic engineering, yea-saying and nay-saying, that polarize the issue. Nay-sayers are obsessed about playing God and lean heavily on slippery-slope arguments, an intellectual fallacy that assumes that one action must inevitably lead to a chain of other more serious and undesirable actions. Such nay-saying he terms actually a legalistic ethics that leads to *ethicism,* with ethicists as experts who decide what is right. Yea-sayers talk about human control and self-realization, but Ladd finds this a cover for a program of consumerism. He feels it also reverses the relation of science and ethics because ethics is not a survey of a ''multitudinous world of possibilities'' but starts in the present with human purposes, needs, responsibilities, and relationships. These determine what we need science for and how we should use it. Ladd then discusses the abundance of myths in both ethics and *scientism.*

The best approach, Ladd suggests, is to try to ''untangle and clarify the underlying issues and conceptual problems.'' We need to question the questions. He then distinguishes problems from issues. Problems are practical and ethical, involve evil, and call for practical solutions. Issues are differences of opinion, disagreements, or controversies. Tay-Sachs disease would be a problem; in vitro fertilization would be an issue. The arguments appropriate for dealing with controversies are different from those for dealing with problems. Ladd also makes an important distinction between macroethics (social problems and social goods) and microethics (individual problems and individual goods). Eugenics is an example of macroethics; genetic counseling of microethics. This resembles the distinction Childs, Hubbard, and Murphy have been working with. He identifies one of the principal problems of macroethics as control (social, political, or economic) and feels this is more crucial than other more spectacular genetic engineering concerns. Ladd stresses that social considerations (macroethics) should not be mixed with considerations determining individual decisions (microethics). He believes that what may be a good reason for an individual may not be a good reason for society and vice versa. He also cautions against working from a rights perspective and attempting to formulate firm rules (rule fetishism), viewing it as a principal source of inhumanity. In place of rules, he suggests principles, such as autonomy or moral integrity.

Howard Brody sums up chapters 2 through 8 in his chapter 9, while suggesting a new analysis of the ethical arguments. He calls attention to the widespread application of genetic engineering techniques and the failure of philosophical understanding to keep pace. However, he feels the fault does not lie with the ethical discipline, since this same period has seen advances in our understanding of other biomedical issues. Ladd's criticism of traditional ethical analysis is more inclusive, however, and not restricted to genetic issues. Brody attempts a comprehensive review of the ethical arguments given in issues of genetic engineering and contends they sort themselves into two categories of pro and con arguments, very reminiscent of Ladd's nay-saying and yea-saying categories. Brody further subdivides these categories, however. The pro side contains a utilitarian argument, a rationality argument, and a genetic-decay argument. The con-side set is a bit longer: the unnatural argument, the argument from uncertainty, the argument from ignorance, the bad-values argument, the future-abuse argument, the resource-allocation argument, and the unethical-means argument.

Brody's next task is to try to narrow this list to analytically manageable proportions. Not all these arguments have equal weight; some are so philosophically weak that they can be easily eliminated. The uncertainty and unnatural arguments are of this sort; they would eliminate *all* behavior if applied. Brody also feels the rationality, genetic-decay (as Murphy also maintained), and future-abuse arguments can be disposed of. The deontological objection to genetic engineering—the unethical-means argument—carries more weight for Brody, especially when framed in terms of respect and personhood, but he finds Paul Ramsey's use of it to prohibit in vitro fertilization faulty. However, since other arguments will carry the important aspects of this argument, Brody prefers to discard it as well. We are left, then, with four arguments that stand up to careful ethical analysis: on the pro side, the utilitarian position (cost-benefit); on the con side, the bad-values argument (violation of social mores), the argument from ignorance (fears of disturbing the complicated natural balance), and the resource-allocation argument (fairer distribution of resources). Since the utilitarian argument stresses the benefits of a cost-benefit analysis and the other three involve the risks or cost, Brody maintains they are all part of the same argument structure, cost-benefit. However, Brody feels it is necessary to add to this a fairness or equity principle, similar to John Fletcher's view, that "as much as possible, risks are borne by those who will stand to gain the benefits." The issues in genetic engineering, therefore, are handled by a modified consequentialist or utilitarian approach, in Brody's analysis.

Though it is impossible to capsulize the overall conclusions from the chapters presented in this part, it would appear that genetic engineering technologies are of recognizable value. The ethical analysts here have concentrated on an approach to bioethics issues based on the formulation of principles: freedom and fairness (John Fletcher), moral integrity (John Ladd), modified cost-benefit (Howard Brody), and the relationship of moral principles to public policy (Warren Reich). Although an approach to ethics based on principles is a common philosophical method (William Frankena, for example, develops a beneficence principle), it is not the only way an ethical theory can be constructed. John Fletcher alluded to Joseph Fletcher's situational approach that stresses a pragmatic methodology and a foundation in feelings or emotion (*agapē*) rather than rationalistically derived maxims and rules. However, it is fair to say that, overall, ethicists have restricted themselves to an ethical rationalism that has not been strenuously critiqued because its assumptions agree with our cultural mores of autonomy, equity, and utilitarian decision making.

2 The Value of Genetic Information

Barton Childs

How useful is information about human genetic variation? How desirable is such knowledge? What is it worth? Value varies with context so I will discuss the usefulness to human beings of genetic knowledge within the context of human biology and medicine. Although I cannot avoid touching on ethical issues, I have assumed that those will be given a workout in other chapters; I will emphasize benefits and costs.

Contexts

It may be useful to distinguish human from medical genetics. To the former, any heritable human variation is grist for its mill; the content of the latter is limited to variation having to do with health. These two aspects of genetics bear the same relation to one another as biochemistry bears to the study of metabolic disease, and like biochemistry, in which fundamental discoveries constantly illuminate disease, today's human genetics is tomorrow's medical genetics.

This translation is usually made in the form of spin-offs of studies undertaken simply to learn something about biological questions. For example, population geneticists are interested in learning how variable populations are and how genes achieve their frequencies—whether by selection or by chance. This argument is far from settled, but important by-products are the discovery of how polymorphic the human gene pool is and the discovery of genes that contribute to variations leading to multifactorial, common diseases.

Other investigators are interested in chromosome structure and the organization and arrangement of the genes therein. Such chromosome linkage is likely to be useful in diagnosis. We already have some hints in the relationship of human-leukocyte-antigen (HLA) haplotypes to autoimmune and other diseases, and there are possibilities for antenatal diagnosis.[1]

Developmental biologists are asking what turns genes on and off and what forces control embryogenesis. Although the yield so far has been

This work was supported in part by a grant from the National Institutes of Health, HD 00486.

limited, we may expect to learn a good deal about the causes (and possibly prevention) of some congenital anomalies.

Then we have recombinant DNA. A few human genes have been more or less isolated, restriction enzymes are being used in mapping, and there is promise of sequencing much of the human genome. Spin-offs promise to be medically (and financially) profitable, and restriction-enzyme methods are already in use in the antenatal diagnosis of sickle cell disease and thalassemia. Gene transplantation, however, is not feasible and so is probably not worth talking about, much less worrying about.

On the medical side, genetic information is being increasingly employed in the pursuit of the conventional medical missions of diagnosis, treatment, prognosis, and prevention. Most of this information has been discovered in the process of the study of patients with familial diseases or anomalies. The rate of its accretion is seen in table 2–1, taken with little change from the fifth edition of McKusick's catalogue.[2] Certainly in the sixth edition, which ought to appear soon, the number listed will be over 3,000. Figure 2–1 shows the exponential increase in the number of inborn errors of metabolism. In these conditions, because we know the primary defect, we can design treatments, some of which are remarkably successful, making reasonably healthy productive people out of those who might have become retarded or deformed. The number of syndromes associated with chromosome abnormality parallels these others, although accumulating at a lesser rate, and the discovery of heritable variations in chromosome structure is proving useful in the linkage game.

As for prevention, genetic counseling, in which reproductive odds and other aspects of genetics are discussed with interested parties, has become a routine part of medical practice; and a growing number of counselors are plying their trade in medical-genetics clinics the world over. Screening tests have been devised to facilitate the early discovery of patients with inborn errors and other disorders, and some of these are being used as a matter of public policy.[3] Testing for heterozygous states as a preliminary to reproductive counseling, as well as for antenatal diagnosis, is also proliferating. Finally, medical genetics has at last begun to extend beyond inborn errors

Table 2–1
Accretion of Mendelian Disorders in Man

Type of Genetic Disorder	*1966*	*1968*	*1971*	*1975*	*1978*
Dominant	837	893	1,043	1,218	1,489
Recessive	531	629	783	947	1,117
X-linked	119	123	150	171	205
Total	1,487	1,645	1,976	2,336	2,811

Source: Victor A. McKusick, *Mendelian Inheritance in Man,* 5th ed. (Baltimore: Johns Hopkins University Press, 1978), p. xiv. Reprinted with permission.

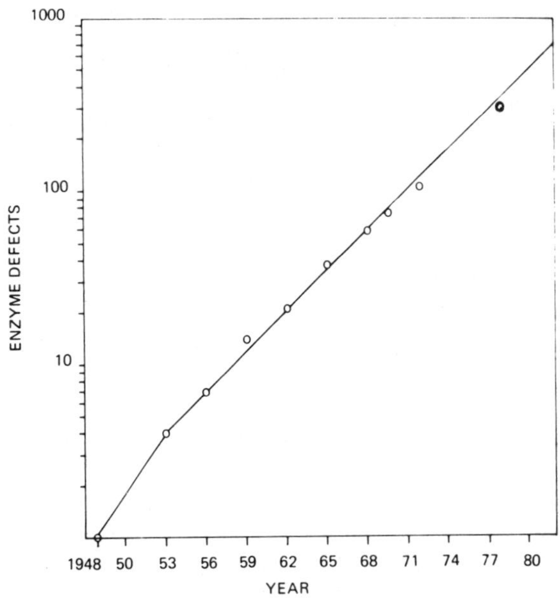

Figure 2–1. Exponential Increase in Number of Inborn Errors of Metabolism

and other single-gene disorders and to be concerned with genetic risk factors, characteristics that appear to be involved in the causes and pathogenesis of the common diseases of middle and late life—for example, high blood pressure, heart attacks, gout, chronic obstructive pulmonary disease, and the like.

I have devoted this much space to this brief outline of human and medical genetics largely to draw attention to the rate at which this knowledge of hereditary human variation is accumulating. I will return to the significance of its present and future mass later.

The Value of the Information

Any new knowledge about genetic variability in any species adds to our understanding of the complex interrelationships of living organisms and is, in my view, worth having. The value of basic science not directed to applied goals is also obvious, although limitation of resources makes necessary the application of priorities. The setting of these priorities is, in the United States anyway, reasonably democratic, so that if something really important is passed over, it is by chance, or because it is before its time, rather than as a result of arbitrary control of funds. The way that kind of knowledge is applied raises questions of benefit and cost, to say nothing of ethics.

I have already pointed out the virtues of the study of nonapplied human genetics, but I can see no legitimate way to use that knowledge of human genetic variation in any but the medical context. The idea of using heritability coefficients as criteria for public policy has been rejected, as it deserves to be, and I know of no information presently available that makes sense of any eugenic aim. So our question is: What gives weight and value to some information and less to other in the medical context?

There is little argument—in principle—over the use of genetic knowledge in the diagnosis, treatment, and prevention of diseases that are treatable and for which the screening and diagnostic tests are good. The first question of value arises over this question of tests. A test must be specific and sensitive—that is, the test must detect only the trait it is supposed to discover and must do so without missing people who have it. We can forgive a moderate weakness on the side of specificity because back-up tests will sort the true positives from the false. However, lack of sensitivity should preclude the use of the test.

Thus if, for example, a child is hemophilic, there is little argument that what we know of genetic transmission should be applied. We should assure ourselves that the mother is or is not a carrier, after which we should give relevant advice. Tests for carriers are greatly improved, although not infallible; errors are on the side of sensitivity. If the mother can be shown to be a carrier, we are also responsible for telling her the odds for her female siblings and for suggesting she discuss the issue with them. She may demur, however, and there is argument over whether or not siblings should be told anyway and who is to do it. There is some question of the legality of telling others if the mother is unwilling.

If a patient has intestinal polyposis, however, it is just good medicine to tell him or her the odds for children and siblings—the disease is a dominant with a bad prognosis—and to add that it is a matter of life or death for them.

If a baby is found by screening to have PKU disease, there is no argument about giving the odds for subsequent pregnancies. Indeed, failure to do so may lead to serious consequences in the way of malpractice suits. I have been teaching genetics for nearly thirty years, and it is humbling to realize that the urge for self-protection is likely to accomplish what hard work and enthusiasm have failed to do.

Thus, we have areas in which the usefulness of knowledge is essentially unchallenged. However, questions begin to arise in those circumstances for which the exact relationship of the genetic quality of the disease, or its treatment, is in doubt. For example, Goldstein and Brown have shown a very clear relationship between a deficiency of a low-density-lipoprotein receptor in the membrane of the cells of some persons and some probability of heart attack.[4] The defect is due to a mutant at an autosomal locus, and homozygotes have a much greater probability of earlier disease, but no one knows exactly what the probability is for the heterozygotes or how surely diet modification will reduce it. So, the question is who should be tested and what to do if the test is positive? Many physicans test only the relatives of people with early heart attacks and recommend moderate diet, exercise, and so forth. However, pediatricians are clamoring to know whether they should test children and, if the tests are positive, whether the children should be restricted to dreary diets and maybe doomed to anxiety too, and uncertainty about the future, and possibly even a sense of personal impairment.

Genes at the HLA locus have been found to be associated with a number of autoimmune diseases. One of these, B27, has been found to be strongly associated with ankylosing spondylitis, a crippling and progressive form of arthritis of the spine.[5] Of white people in the United States, 8 percent have the B27 haplotype, and one-fifth of these are said eventually to develop the disease, while 80 percent of cases of ankylosing spondylitis have it. Here we have very serious questions of specificity and sensitivity. Twenty percent of cases do not have B27, and four-fifths of people with B27 never develop the disease. Thus, this particular piece of genetic information is essentially useless for diagnosis. Still, it is very valuable as information in regard to cause, since the HLA association may be telling us something about a linked immune-response gene that is involved in cause or pathogenesis, or both. So the information is useful in general, useless in specific. It has been abused, however, and the medical literature tells of B27 cripples—people who do not have the disease but who have B27 and some other form of backache.[6]

The questions mount when we do tests of heterozygotes. These are healthy people, and the tests have to do only with reproductive advice. Here the benefits can be overwhelmed by costs unless:

The people being tested are thoroughly informed as to the nature of the disease, what their risks are, and the potential to them of being tested;

The test is done at an appropriate time, usually when a couple wants to reproduce or at least at a time when a person can handle possibly threatening information;

The testing is done in ways and places that preserve and protect confidentiality;

Tests are of good specificity and sensitivity;

Adequate resources exist for backup tests, definitive treatment, and counseling.

If there are holes in this matrix of conditions, the testing can do more harm than good, but if all are met, this qualifies as good medical care. Most of these conditions were not met ten years ago when sickle cell carrier screening was begun, and some harm was done, but most of the conditions have been met in the widespread Tay-Sachs screening programs.

A few years ago, several colleagues and I studied 739 participants in the first Tay-Sachs screening program organized by Kaback in the Baltimore-Washington area.[7] Included were 120 persons found to be carriers and their noncarrier spouses. Among many other things, we were interested in how the carriers—and their spouses—felt when they learned that they had the gene. Just under one-half of the carriers said they were shocked, angry, or felt in some way compromised or imperfect. Time and counseling, however, erased these feelings, and by the time we interviewed them some months later they were reconciled and enthusiastic about the program; indeed, 94 percent said they would do it again and were relieved to know they were not eligible to have a Tay-Sachs baby. I know of no studies to prove the point, but I believe these feelings of diminution and apartness are likely to be minimized when such testing becomes simply a part of routine preventive medicine and when people are aware that all of us have some "imperfections" in our genotypes.

The classical case of cost-heavy information would be a specific and sensitive test for the Huntington's disease gene in at-risk people. Those discovered to have the gene would know they were doomed and had no escape. That seems to me too great a cost, but there are no less than three reports of studies of persons at risk that tell us that 77 to 84 percent of these people believe that such a test should be available and that they might be willing to be tested.[8] What such people would do if actually confronted with a test, and how they would respond to being told they have the gene, is unknown, but a good number are giving it thought, and I have no doubt that such a test will be forthcoming soon.

The evaluation of benefits and costs is particularly difficult when knowledge of genetics is used in antenatal diagnosis. This is an entirely new form of preventive medicine in which the disease is prevented by forestalling, selectively, the birth of an affected person. The benefits are easily recounted: removal of the threat of a heavy burden for parents and relief of a financial burden ordinarily borne by society. The evaluation of the life that never is is more difficult. Some studies of people with neural-tube defects suggest that being—that is, living—with burdens is preferable to not being.[9] Also, it is usually not possible to know whether the fetus that is aborted would have been severely or mildly affected.

The costs of this procedure are the life of the aborted fetus and the feelings of the parents who participated in the selective abortion. To some, the former cost is too great to allow the procedure. That is a defensible view, although it seems to me that as a corollary it should carry a very strong sense of responsibility for the hapless victims of the not-prevented diseases.

Deplorably too few studies have been done on the psychological and social impact on the lives of parents who have participated in selective abortion. Studies that do exist tell us that some, at least, pay the price in depression and dislocation.[10] For those people, however, it is not too high a price, since some did it again and very few regretted it. Still, much more study is needed to learn who is likely to suffer these effects and how, if there is a way, to avert them.

Conclusion

In contemplating the uses of genetics in medicine, I am struck by two things: (1) that benefits can be maximized and costs can be minimized by the choice of conditions under which the knowledge and techniques are applied and (2) that human variability counts for much here as it does in all aspects of living; the benefit-cost equation is weighted differently for different people. What is too costly for one may be perceived as beneficial by another. So it seems to me that a test of the sincerity of our vaunted pride in pluralism, individuality, and freedom of belief lies in our readiness to support the expansion of genetic knowledge and its use in broadening options for all. The word *options* here is crucial. To exercise an option, one has to be able both to recognize it and to use it in making a considered decision rather than to be carried along on a tide of fashion or to be patronized by do-gooders. That means emphasis on education in human biology and genetics at all levels, and I am not much encouraged by the mentality of a government that closes down the efforts of the National Science Foundation to do just that.

Finally, I return to the rate of the accretion of our knowledge and its present and future mass. We seem to be accumulating markers for susceptibility to diseases—we are making lists of human vulnerabilities. They do not say that you are sure to have this or that disease, or that if you lack them that you will escape, they simply pose probabilities, but they are probabilities that, for some, may raise anxieties unimagined in the state of ignorance. We know only a few of such diseases now, but cast your mind forward to a time when we will know dozens, even hundreds. Probably we are all susceptible in some degree to some unpleasant things, some of us to many. For some of these diseases we will undoubtedly find preventive measures, but what if there are none? In any case, technological advances seem always to create new problems, even while solving old ones, and to bring new uncertainties to living. How we handle the uncertainties that will follow this expanding knowledge of human and medical genetics will pose the most formidable test of its value.

Notes

1. D. Teichler-Zallen and R.A. Doherty, ''Amniotic Fluid Secretor Typing: Validation for Use in Prenatal Prediction of Myotonic Dystrophy,'' *Clinical Gentics* 18 (1980):257–267.

2. V.A. McKusick, *Mendelian Inheritance in Man,* 5th ed. (Baltimore: Johns Hopkins University Press, 1978).

3. *Genetic Screening; Programs, Principles, and Research* (Washington, D.C.: National Academy of Sciences, 1975).

4. M. Brown and J. Goldstein, ''Familial Hypercholesterolemia: Model for Genetic Receptor Disease,'' *Harvey Lectures* 73 (1979): 163–202.

5. A. Calin, ''HLA B27: To Type or Not to Type?'' *Annals of Internal Medicine* 92 (1980):208–211.

6. Ibid.

7. B. Childs; L. Gordis; M.M. Kaback; and H.H. Kazazian, Jr., ''Tay-Sachs Screening: Social and Psychological Impact,'' *American Journal of Human Genetics* 28 (1976):550–558.

8. R. Stern and R. Eldridge, ''Attitudes of Patients and Their Relatives to Huntington's Disease,'' *Journal of Medical Genetics* 12 (1975):217–233; J. Barette and C.D. Marsden, ''Attitudes of Families to Some Aspects of Huntington's Chorea,'' *Psychological Medicine* 9 (1979):327–336; and B. Teltscher and S. Polgar, ''Objective Knowledge about Huntington's Disease and Attitudes towards Predictive Tests of Persons at Risk,'' *Journal of Medical Genetics* 18 (1981):31–39.

9. S. Dorner, ''Adolescents with Spina-Bifida. How They See Their Situation,'' *Archives of Diseases in Childhood* 51 (1976):439–444.

10. J. Robinson; K. Tennes; and A. Robinson, ''Amniocentesis: Its Impact on Mothers and Infants. A 1-Year Follow-up Study,'' *Clinical Genetics* 8 (1975): 97–106; B.D. Blumberg; M.S. Golbus; and K.H. Hanson, ''The Psychological Sequelae of Abortion Performed for a Genetic Indication,'' *American Journal of Obstetrics and Gynecology* 122 (1975):799–808; and P. Donnai; N. Charles; and R. Harris, ''Attitudes of Patients after 'Genetic' Termination of Pregnancy.'' *British Medical Journal* 282 (1981):621–622.

3 Some Practical and Ethical Constraints on Genetic Decisions about Childbearing

Ruth Hubbard

I begin this chapter by putting this topic in its historical context and asking why we are talking about these kinds of genetic decisions and why at this time. Throughout recorded history, people have had questions about heredity—where our traits come from and why children resemble their parents. During the nineteenth century, however, attention to heredity became important for political and social reasons. Hereditarian ideas came to be essential ingredients of nineteenth-century Euro-American liberalism—of political notions of equality, equal opportunity, and meritocracy. The belief that these ideals were basic to our political institutions required that one find acceptable explanations for the existence of obvious, and often increasing, inequalities between different groups of people—rich and poor, blacks and whites, women and men. How are we to reconcile obvious inequalities in economic status and political power with pretenses of equality? The ready answer is that social inequalities between groups are an expression of inborn inequalities between individuals. Since everybody knows that people differ, this easy and unspoken transition from the traits of individuals to group characteristics carried a logic that was readily accepted into the basic assumptions of nineteenth- (and indeed twentieth-) century liberalism.

It is important to recognize that this explanation—that our biological heredity determines the position each of us occupies in this society—was embedded in nineteenth-century Western social thinking well before there was what we would now call a science of genetics—that is, before the rediscovery of Mendel's work around 1900, which marks the beginning of modern genetics. For example, in 1869 Francis Galton, upper-class Briton, cousin of Charles Darwin, mathematician, and the man who coined the word *eugenics* and started the eugenics movement, wrote a book he called *Hereditary Genius*. In it he set himself the task of explaining why it is that getting honors at Cambridge University runs in families, as does becoming a judge, statesman, divine, literary man, or scientist. And he claimed to be able to demonstrate that these capacities were biologically inherited and that Britons were particularly well endowed in these regards. You can judge the objectivity of his approach from his statement that "The natural ability of which this book mainly treats is such as a modern European possesses in a much greater average share than the men of the lower races.'' (And note, that this pattern of inheritance operates only in the male line!)

The eugenics movement that Galton initiated was a political effort to better the human race by encouraging those from the ''better stocks'' to produce more children (called positive eugenics), while restraining or preventing the ''less desirable'' elements in the population from reproducing (negative eugenics). (Among less desirables were included the mentally ill and feebleminded as well as habitual paupers and criminals.) This was a logical extension of the belief that each of us individually harbors the causes for why things go well or badly with society—hence consistent with nineteenth-century liberalism and individualism—and it led to the conclusion that it is morally and socially incumbent on scientists and legislators to do what they can to improve the human stock biologically.

Now, eugenics is a term that has hardly been used during the last several decades since it fell into disrepute in the early 1940s because of what happened in Nazi Germany, but it is beginning to be used again. Therefore, I want to tell you a little about it because eugenics had significant deleterious social consequences in this country as well as in Germany. The word *eugenics* means ''race improvement.'' The eugenics movement was sponsored and led by people, many of whom were not themselves geneticists but who used genetic or pseudogenetic arguments to underwrite racist social policies. Meanwhile, though many practicing geneticists supported the movement in its beginnings, a majority eventually were put off by its frankly conservative politics and explicit racism. However, though most geneticists after about 1920 came to believe that scientific arguments were being misused to pretend that we can ''improve'' the human race through selective-breeding practices, few of them spoke out against the eugenics movement. Thus it became possible for eugenicists and other scientific racists to adopt, without vocal scientific opposition, what the British biologist Steven Rose has referred to as ''a tone of pious resignation, the regrettable acceptance of biological laws of human inequality, by hard-headed realists with the self-appointed task of biological law enforcement.''[1]

The most important political victory of the U.S. eugenics movement was the passage of the Immigration Registration Act in 1924. Though ostensibly intended to limit immigration in general, it had strong and explicitly racist components in that it was designed to decrease the representation of eastern and southern Europeans (immigration from Asia had been curtailed earlier) so as to give greater predominance to Nordics—that is, Americans of British and north-European origins. This was achieved by restricting the annual number of immigrants permitted to enter from each European nation to at most 2 percent of U.S. residents listed in the 1890 census (this in 1924) who had been born in that country. The date, 1890, was selected deliberately to establish as a baseline the composition of the U.S. population prior to the arrival of large numbers of immigrants from eastern and southern Europe around the turn of the century.[2] One disastrous, later result of this discriminatory policy was to prevent entry into the United States of Jews of east-European origin as well as of other eastern Europeans who were trying to flee the Nazis after 1933 and who subsequently were killed in German concentration camps.

The other major "success" of the U.S. eugenics movement was the passage of eugenic sterilization laws. A statement by Lewis Terman, one of the chief engineers of U.S. IQ testing, represents the underlying rationale:

> The fecundity of the family stocks from which our most gifted children come appears to be definitely on the wane. . . . It has been figured that if the present differential birth rate continues 1,000 Harvard graduates will, at the end of 200 years, have but 56 descendants, while in the same period 1,000 S. Italians will have multiplied to 100,000.[3]

To avoid such a fate, by 1931 some thirty states had passed compulsory sterilization laws. Though most of them largely went unenforced, by January 1935, they nonetheless had resulted in some 20,000 forced sterilizations, nearly half of them in California. In general, aimed at the feebleminded (broadly interpreted), "some of these laws applied to a very wide range of 'hereditary defectives,' including 'sexual perverts,' 'drug fiends,' 'drunkards,' 'epileptics,' and 'diseased and degenerate persons.' "[4] These laws also had obvious class and racial biases, at least in the ways in which they were enforced. As Lancelot Hogben phrased it: "There is a strain of haemophilia in the Royal Houses of Europe. No eugenist [*sic*] has publicly proposed sterilization as a remedy for defective kingship."[5] The spirit and application of these laws were not very different from present-day practices whereby poor women, particularly blacks, Hispanics, and native Americans, are sterilized against their wills under the banner of population control.[6]

In Germany, a society for racial hygiene was founded in 1902, at about the same time that the eugenics movement became a social force in the United States. In 1920—thirteen years before Hitler's rise to power—a German jurist and a psychiatrist published a book called, in translation, *The Release and Destruction of Lives Devoid of Value,* which laid out the theory behind what became the German eugenic policies. Everything was ready to enact in 1933, the very year Hitler became chancellor, a eugenic sterilization law that established so-called hereditary-health courts. During the first year of its existence, 56,244 people were judged hereditarily defective and were sterilized. Directed initially against the insane and feebleminded, the policies later were extended to homosexuals and other "undesirables."

Eugenic measures in Germany escalated from sterilization to euthanasia and eventually led, as everyone knows, to the murder of millions of Jews, gypsies, and other social or political outcasts.[7] These various excesses put eugenics in disrepute, but it is important to remember its history and social and scientific origins in Britain and the United States as well as in Germany.

There is another reason why the enthusiasm for negative eugenics essentially had disappeared by the end of World War II—that is, even when shorn of its discriminatory and racist intent and limited in its application to traits that can be shown to be inherited biologially, such as diseases like PKU (phenylketonuria) or sickle cell anemia, negative eugenics does not work. To prevent people who have such a disease from reproducing does not appreciably reduce the incidence of the disease in the population. The reasons for this were stated clearly in 1908, in what geneticists call the Hardy-Weinberg Law. It is interesting that it was not until the 1940s—more than thirty years after its scientific basis had been eroded—that the pseudoscience of eugenics was stemmed.

The point is that most ''deleterious traits,''[8] such as PKU or Tay-Sachs disease (a degenerative, neurological disease of young children), are what geneticists call recessive. What this means is that if a person has a gene for this trait on only one chromosome, rather than on both members of that particular chromosome pair, the trait remains hidden—that is, the ''defective'' gene recedes behind the normal, functional gene that is carried on the other member of the chromosome pair. People who carry one recessive ''abnormal'' gene and one ''normal'' one on a pair of chromosomes usually have no way of knowing about it because, to a first approximation, that trait appears to be normal. One must usually perform genetic or biochemical tests to find out that these people are carriers of the recessive gene (such as, for example, the gene for Tay-Sachs disease or for sickle cell anemia). The trait itself—the ''disease''—only appears in an offspring if, by chance, two people who are carriers for the same recessive gene have children, in which case each of their children has one chance in four of exhibiting the trait. The appearance of an individual with a recessive genetic disease—that is, one who has two genes for that trait—is a fairly rare event in a population and one whose likelihood can be described only statistically, though some recessive diseases occur more frequently in some racial or ethnic groups than in others. (For example, 158 children with Tay-Sachs disease are born for every million births among Ashkenazi Jews, whereas the incidence is only 1.7 per million births among people of non-Ashkenazi descent.)

The reason why preventing people who have a recessive inherited disease from having children does not appreciably decrease the incidence of the gene in the population is that almost all the genes for that kind of disease are carried singly (and not in pairs) by people who do not know it;

and they have no way of knowing it unless we were to test everyone for all possible deleterious recessive traits—something that is neither feasible medically or economically nor socially desirable. To say it another way, people who actually exhibit this kind of genetic "disease" are only the very tip of the iceberg, while the overwhelming majority of recessive genes lies submerged in the population. To prevent people who have a "genetic defect" from having children may help them personally, if they do not want to have affected children, but it does not appreciably decrease the frequency with which the recessive gene occurs in the population.

In the real world, the genetic game of chance means that although each of us carries recessive genes for between five and eight seriously deleterious traits, and many more for milder defects, only about 3 percent of children are born with a serious defect. In other words, any one of us has every reason to expect to have healthy, normal children unless we have a specific reason to believe otherwise. True, occasionally two people who have no reason to anticipate problems have a child with a serious defect or disease. But, this is a risk one has to accept if one wants to have children—and one no greater than many other risks we take in our everyday lives.

However, we have entered a new era in which it is becoming possible to predict with greater accuracy than before the carrier status of prospective parents for some recessive traits and, by means of amniocentesis, to determine whether a particular fetus has any one of a number of specific disorders. Furthermore, the number of disorders that can be diagnosed in these ways is increasing rapidly. In looking at the ethical issues involved in the newer techniques of genetic diagnosis, I want to draw a distinction between two kinds of motivation for perfecting and disseminating these procedures—motivations that often blend into one another in writings by their advocates. One is to help prospective parents who have a particular genetic disease or who have reason to suspect that their future child might be born with such a disease. The other is to rid the human stock or gene pool of genetic defects—that is, negative eugenics. The former clearly is a legitimate concern of the biomedical profession, though one that we as potential "patients" must have the opportunity to evaluate. The latter must be a matter of public policy, and in its development and implementation, self-appointed agents of biological law enforcement (to quote Steven Rose) cannot be given special powers to interpret the laws of nature for us. Unfortunately, physicians often gloss over this distinction and see their role not only as helpers of people who need their services but also as agents of "society's interests" in eradicating hereditary diseases. The risks inherent in this view of the role of biomedicine should be obvious from the brief history of eugenics I have given.

My main quarrel with medical genetics is over this blurring of issues. However, I also have reservations about genetic counseling and amniocentesis at the more personal doctor-patient level. Before I discuss them, I must state explicitly that my position derives in some measure from my autobiography, since I am the mother of two children who were born when I was 35 and 37 years old, and an Ashkenazi Jew whose partner also is an Ashkenazi Jew. If I were planning my pregnancies now, we would be advised to find out whether we are carriers for Tay-Sachs disease. Because of my age, I would also be advised to have amniocentesis to determine whether the fetus has Down's syndrome. However, since my pregnancies happened more than twenty years ago, I escaped having to dwell on their possible pathological outcome. Needless to say, I realized full well that having children—and particularly at my age—involves risks; but I was prepared to take them *because they are small.*

What were my actual risks? By present estimates, at my age the risk of having a child with Down's syndrome was about one in three or four hundred; the risk of a child with Tay-Sachs disease for two Ashkenazi Jews (randomly chosen) is about one in six thousand.[9] If I or my partner had had relatives with either of these problems, the risks would, of course, have been much greater. There would, therefore, have been every reason to explore the situation further and perhaps to forgo having chidren if the risks were too great and if we did not want to take them.

To put these risks into perspective, it is well to realize that in the United States at present, the risk of any one of us being killed in a motor-vehicle accident is about one in five thousand and considerably greater in some localities. Yet we walk the streets and drive on the highways. My or my daughter's risk of being raped is much greater; yet we live in Cambridge, Massachusetts, where the risk is one of the highest in the country. To these we must add the risks from exposure to environmental carcinogens at home, outdoors, or at work and the unknown, though probably formidable, risks of nuclear fallout, accidents, or war. How should we evaluate the various risks we run while living in this society? What personal and political actions make sense? Why do we emphasize the risks we carry in our genes, and why at this time?

Perhaps the reason we are in a period of revived interest in our heredity is that people in this society feel increasingly threatened and insecure about many of the things that go on around them. It is almost as though we gain reassurance from concentrating on the problems inside, rather than always worrying about the seemingly insurmountable problems in the world. And it certainly is in the interest of those in power to point the accusing finger at us and our genes rather than at the many environmental risks to our biological and social well-being. Similar to what happened in the nineteenth century, they now tell us that, if there are too many criminals, do not look at social inequalities; try to pinpoint genes (or chromosomes) for criminality. If increasing numbers of people are getting cancer, do not enact and enforce tighter industrial standards and antipollution measures, but try to find genes that make people sensitive to environmental carcinogens, and so on. This is part of the common tactic of blaming the victim. It has the result that a great deal of energy and trust is put into developing knowledge and technologies to try to change what goes on inside us; all the while, most of us know full well that the major hazards, in fact, are out there, but we do not know how to tackle them. What makes the situation more problematic is that some people indeed run greater-than-average risks of bearing children with inborn diseases, and for them the new technologies may be useful and often lifesaving. The question is where to draw the line between helping those who need professional help and turning all of us who come to believe that we need professional advice before we can undertake to have children into invalids.

I have little doubt that were I to have children now, under the same personal contingencies as existed when I had them some twenty years ago, we would probably decide to be tested for the Tay-Sachs gene (after all, it is very easy), and I would probably have amniocentesis (how can I inflict a child with Down's syndrome on myself, my partner, my other child, my parents, or indeed society when I have the choice not to?). I am only grateful that in my day I did not have that choice. That is one of the points I want you to think about: that as these choices become available, they rapidly become compulsions to choose the socially endorsed alternative. In this realm, it is amazing how quickly so-called options are transformed into obligations that, in fact, deprive us of choice.

The risks of prenatal counseling and amniocentesis probably are mainly social and psychological; hence we cannot attach numbers to them. (A warning: It is still too soon to discount possible long-term hazards from the ultrasound devices that are being used in order to decrease the risks of physical injury to the fetus or placenta during amniocentesis, and that are now rapidly coming into widespread use during routine prenatal examinations.) Risks I have in mind are, first, exaggerating our concern with genetic risks, while neglecting or downgrading the all too prevalent risks in our environments and life situations (including risks from prenatal interventions). After all, we live in the Western nation with the highest infant and newborn mortality rates. This not because of genetic defects but because, among industrialized countries, the United States has the greatest proportions of people living in poverty. Are we really addressing the most pressing medical and social needs when we develop new and more-refined techniques of prenatal diagnosis?

My second concern involves the risk of increasing our dependency on so-called experts to guide us through the important decisions in our lives. One reason I reviewed the history of eugenics was to show how, in this realm, so-called professional opinion and objectivity often conceal ethnic or class prejudices and how easily professional advice shades over into professional (and sometimes also political) control. Physicians and other trained counselors often claim that they only present the facts and do not advise clients what to do. But that is just about impossible if they have strong feelings one way or the other. For example, when both parents carry the same recessive gene do you tell them that they have one chance in four of having a diseased child or three out of four chances of having a normal one? Objectively, these statements are equivalent. At the human level, they are very different; and in many situations, value judgments can intrude much more subtly and often unconsciously.

Let me give you an example of how easily coercion creeps in. A couple of my acquaintance made an appointment for genetic counseling in order for the woman to decide whether she should have amniocentesis since she was 37 years old. When the time came, she found she needed to postpone the appointment by a few days. On the phone, she encountered a distressed receptionist: "But that is when we have scheduled the amniocentesis!" My friend was upset, because until that moment she had assumed that she was entering a two-step process in which she would decide whether to undergo amniocentesis, and only after the counseling session.

My third fear, and it is closely connected to the other two, is that this kind of medicalization and increased dependency on professionals lures women into the false hope that, with proper medical advice and supervision, we can have risk-free pregnancies and problem-free children. When the first problem does arise, as it will inevitably and all too soon, we search for a professional to deal with it. Given the scariness of launching unpredictable children into an unpredictable world, we are liable to fall for technological fixes that convey an illusion of certainty. The availability of not-all-that-objective counselors with their not-all-that-adequate tests serves to increase parents' (and particularly mothers') already excessive and unwarranted guilt. If their child has problems, what did they do wrong? Should they have gone to the genetic counselor before it was conceived as well as to the psychiatric counselor afterwards?

I am not being frivolous. In our society, women tend to be blamed for whatever goes wrong with our children. Yet, at every turn, our confidence in our abilities to handle the very real responsibilities of childbearing and rearing is undermined by the existence of experts whose opinions usually have been formulated in the safety of their studies or offices and are not likely to be tested by the immediacy of personal experience. I see genetic counselors as one more layer of such experts to be added to the obstetricians, pediatricians, psychiatric counselors, social workers, teachers, and guidance counselors, whose expertise often only serves to clothe prevailing societal norms and prejudices in a mantle of professional language and behavior.

Do not misunderstand me. *I support unequivocally a woman's right to choose whether to bear a child and when.* This means that she must have the option not to have a child when she does not want one, whatever her reasons. I also believe, however, that if a woman decides to have a child, she is better off facing the fact that this decision involves risks—many of them unpredictable—and that no matter how careful she is, there will be crises, unhappy times, and problems. I also believe that in the totality of risks for most people, the risks of excessive medicalization and professional control—though they cannot be expressed in numbers—are as real as the risks of abnormality or disease. (Clearly this is not true for people who have personal or familial reasons to anticipate health problems.)

Let us remember that when it comes to interventions in pregnancy and childbirth, the record of the biomedical profession has been mixed at best.[10] They have brought us so-called childbed fever, newborns' cataracts and blindness (retrolental fibroplasia), thalidomide babies, DES (diethylstilbestrol) cancers, and a long list of other medical and psychosocial problems, not least among them the inability of many women to breast-feed their babies.

One last point I would like to make is specifically about Down's syndrome. The range of expression of this set of symptoms is enormous, and the degree of disability can depend a great deal on the child's living situation. Many families cannot undertake to raise a child with Down's syndrome; others believe they can and would rather try than abort a fetus with Down's syndrome. We must be careful not to let personal and societal prejudices against Down's syndrome curtail the rights of families who would rather raise such children than abort them. (Needless to say, I think it outrageous to suggest that women be forced to bear them by removing or curtailing our rights to abortion, as has been suggested by members of Congress, the president, and cabinet officers.) This is particularly important in view of the fact, that, even if all pregnant women over 35 had amniocentesis and decided to abort all fetuses with Down's syndrome, the number of births of Down's syndrome infants would be reduced by only about a third. The reason is that, though the incidence of the chromosome abnormality associated with Down's syndrome (trisomy 21) increases markedly in women over 35, women below that age bear the great majority of children. Therefore, though women under 35 are much less likely to have a child with Down's syndrome than older women, the younger women in fact bear the majority of Down's syndrome children. Thus, Down's syndrome can only be eliminated if all pregnant women were required to undergo amniocentesis and to abort all fetuses with trisomy 21—something few people would suggest or countenance and for many different reasons. However, that means that it is extremely important that we not increase the prejudice and essentially punitive stance against families who have children with

Down's syndrome. While we must support all women (and families) who do not want to have such a child if they can avoid it, we also need to turn our technical and social skills to improving the lives of children with Down's syndrome and of their parents. Parents must not be made to feel that it is their fault or that they should have known better, since more often than not, they cannot know. People will go right on having children with defects. At least as urgent as to forestall this must be efforts to improve the help and services available to the children and their families so that they can cope better, or at least more easily, with their often very great problems.

Notes

1. Steven Rose, "Scientific Racism and Ideology: The IQ Racket from Galton to Jensen," in *Ideology of/in the Natural Sciences,* edited by Hilary Rose and Steven Rose, p. 111 (Cambridge, Mass.: Schenkman Publishing Company, 1980).

2. The effects of the eugenics movement on U.S. immigration policies are discussed in Kenneth M. Ludmerer, *Genetics and American Society: A Historical Appraisal* (Baltimore: Johns Hopkins University Press, 1972).

3. Lewis Terman, "The Conservation of Talent," *School and Society,* no. 483 (29 March 1924):363, quoted in reference 1.

4. Ludmerer, *Genetics and American Society,* p. 92.

5. Lancelot Hogben, *Science for the Citizen* (New York: W.W. Norton, 1938), p. 1053. Cited in Ludmerer, *Genetics and American Society,* p. 93.

6. For a description of present-day involuntary sterilization practices in the United States, see Helen Rodriguez-Trias, "Sterilization Abuse," in *Science and Liberation,* edited by Rita Arditti; Pat Brennan; and Steve Cavrak, pp. 113–127 (Boston: South End Press, 1980). A recent Associated Press dispatch (*Boston Globe,* 17 July 1981) reported that "poor Medicaid recipients" are being sterilized without proper enforcement of federally mandated consent procedures. The article states that "110,000 sterilizations, including 31,000 hysterectomies, were performed in 1980."

7. The history of eugenic practices in the United States and Germany is discussed in Allen Chase, *The Legacy of Malthus: The Social Costs of the New Scientific Racism* (New York: Alfred A. Knopf, 1975). A well-documented account of the early history and escalation of German race hygiene has recently been made available in English in Stephan Chorover, *From Genesis to Genocide: The Meaning of Human Nature and the Power of Behavioral Control* (Cambridge, Mass.: MIT Press, 1980).

8.. I stress the fact here that these labels depend on social conventions as well as on scientific description, and often largely on the former.

9. These numbers are taken from A. Milunsky, J.W. Littlefield, J.N. Kaufer, E.H. Kolodny, V.E. Shih, and L. Atkins, "Prenatal Genetic Diagnosis," *New England Journal of Medicine* 283 (1970):1370–1381, 1441–1447, 1498–1504.

10. Comprehensive and well-documented criticisms of such interventions have recently been published in Helen B. Holmes, Betty B. Hoskins, and Michael Gross, eds., *Birth Control and Controlling Birth: Women-Centered Perspectives* (Clifton, N.J.: Humana Press, 1980); and in Holmes, Hoskins, and Gross, eds., *The Custom-Made Child? Women-Centered Perspectives* (Clifton, N.J.: Humana Press, 1980).

4 Genetic Engineering: Eugenic Consequences

Edmond A. Murphy

At the outset, let me say that I intend not to prescribe but to dissect a problem. I believe that the major obstacles to devising an appropriate course of action are not ethical in any technical sense, but rather involve the difficulty in coherently assembling the facts and the issues. I shall attempt to do so from the standpoint of the clinical and mathematical geneticist. I shall at least try to exhibit their interface with values. Beyond that, you must piccc out my imperfections with your own thoughts.

To begin with, I do not at all agree with the ancient nihilism that one cannot deduce an obligation from a fact, a nihilism that seems to me to stem from ambiguities surrounding the nature and meaning of obligation. For my part I shall be content to engage in enthymemes. Suppose I say to a patient (as I do, on occasion), "If you do not take your insulin, you will die within a month." Formally speaking, I have exhibited no obligation, merely fact. Whether or not he should want to die is a separate issue. In certain circumstances it, too, may be deduced from fact. For if the patient wishes to live to see the twenty-first century, clearly he is obliged not to die in 1979, and therefore he is obliged to take his insulin in 1979. Even if, in the ultimate economy of things, it were proper for me to arrogate to myself the decision as to what the patient ought to want, there are many more immediate pressures and obstacles to my doing so: the specters of paternalism and bigotry, the protection of the patient by the law, the inviolable sanctum of conscience, and much more. So I do not propose here to lay out a blueprint for society; that has been done by others both much cleverer and much stupider than me. I will have more than enough to do in grappling with my own area of supposed competence. I imagine that there will be no lack of people who will supply the missing premise in the enthymeme and complete the syllogism.

My second preliminary point arises from a subtle variation on an old theme. If you read the more sensible critics of evolutionary theory, let us say between the two world wars, you will encounter a common argument that runs something like this: The evolutionist says that natural selection eliminates the bad variations and conserves the good so that there is a trend

The work in this chapter was supported by NIH Grant #1-RO1-GM 24736.

to improvement. But telling us that selection operates thus does not tell us about the origin of the variation, for a sustained evolution requires a sustained source of variation. Moreover, the only readily verifiable illustrations of selection that the biologist has produced are the instances in which a harmful type is eliminated. This model tells us where bad variation goes to, but not where good variation comes from.

Like all arguments from lack of knowledge, this criticism has an ephemeral and dated character, and the vast insight that we have gained in the last thirty years into the nature of mutation will seem to have allayed the uneasiness their criticism should have produced in the unprejudiced mind. However, sometimes critics are wiser than they know, and we are just beginning to see some disquieting cracks appearing in the smooth facade of evolutionary orthodoxy. One feature is the nascent realization that the core of the old skepticism, although wrongly focused, is sound—that is, that the process by which genes are perpetuated is not simply the converse of the process whereby they are eliminated. A genetic stock with an average of less than one progeny per parent is certain to become extinct, but the converse does not hold. Indeed, I have had recourse to showing explicitly, for the benefit of the doubters, that average size of family may be as large as you please, generation after generation, and yet extinction remain almost certain, while those lines with small average family size may yet be certain to survive.[1] The grocery-store arithmetic of yesteryear is not good enough; neither, for that matter, is the more sophisticated, but still naive, stochastic theory by which this last-mentioned paradox has been proved. A proper treatment, which will be vastly more difficult, would incorporate feedback mechanisms, random variation in the environment, and the pyrotechnical complexities of the stochastic theory of ecological interaction. We are not yet quite ready to carve evolutionary theory in marble.

But there is another ugly, great crack appearing in classical genetics that promises to work even greater havoc. Most of the explicit studies of genetic fitness have been based on harmful mutations. Let us say, as a simple illustration, the dominant disorder, Marfan syndrome, appears out of the blue, in all likelihood by new mutation. It is a disorder serious enough to interfere with reproduction, and while there *may* be affected descendants, they will not be numerous. Accordingly, any particular mutant line will usually die out in two or three generations. In the casual encounters of the clinic, big pedigrees will, because of ascertainment bias, tend to be overrepresented. In spite of this fact, we rarely see big pedigrees in Marfan syndrome, and with Pauline boastfulness, I would claim that at Johns Hopkins we have unrivalled experience of this condition.[2]

Now there are two facets to this illustration worthy of further comment. First, it is easy to show empirically that clinically deleterious mutations are unlikely to survive because we know the final outcome in a few generations. However, to show that a beneficial mutation will persist indefinitely takes, naturally, an indefinitely long time; and I am sure my grant will have run out before I attain cogent proof. Indeed, I shall certainly have run out myself. Thus the positive aspect of evolution is, in the nature of things, difficult to demonstrate empirically, and while some enthusiasts have tried a cut-price retrospective proof, a little serious thought will show that, by so doing, they have stepped into the quagmire of an old tautology: All they are proving is the survival of the survivors. The fundamental problem (seen epidemiologically) is that they have a numerator but no denominator, a sample with no reference set.

The second comment is vastly disturbing: that, contrary to optimistic argument of long standing, the catastrophic mutations tell us nothing about evolution and very little about normal selection. Manifestly, they tell us nothing about evolution because a mutation that is certain to become extinct naturally has no permanent impact on the gene pool except indirectly and *per accidens*. Further, we are not even clear that it throws any light on its normal counterpart except in a trivial sense. As an analogy, let us suppose that I have a television set. Some noble cause inspires a group of zealots to burn it. Now what can we infer from the fact that the television set promptly ceases to work? I suppose we can say that its normal functioning depends on an absence of fires, as also of earthquakes, bombs, and other physical catastrophies; but I am not satisfied that research on design for television sets should center around the study of fires, earthquakes, and bombs. Such massive, but ontogenically irrelevant, changes I have elsewhere called disruptive interlopers.[3]

Now there is a mounting suspicion that, in the same kind of way, much of classical population and quantitative genetics is beside the point. On the one hand, several lines of enquiry suggest that the gene is not the primary unit of evolution at all, in higher organisms at least. I may cite such disparate lines of evidence as Stanley's analysis of the fossil record,[4] Boyer's studies of the genetic load in wild populations of monkeys,[5] and his work with Smith, Kunkel, and others on reiterated DNA.[6] On the other hand, I may instance the logically bewildering behavior of systems involving two genetic loci or more. In quantitative genetics, there is a just perceptible ground swell of opinion[3, 7] that the days of the Athenian democracy—the theory that multiple effects of the several loci and of the environment are more or less equal, independent, and linearly additive[8]—are numbered. To keep the subjects of quantitative genetics and evolution alive, we are going to need more than a tissue of superficial tautologies, notionally unilluminating and empirically incorrigible. Even if it *is* true that the Athenian democracy does apply to neutral traits, we now know far too much about the biochemistry and physiology of feedback mechanisms, redundancies, competing pathways, serial metabolic processes and cascades, and much more to believe that disease could possibly be analyzed in naive Galtonian paradigms.

I have thought a good deal about this area of genetic engineering and find I know almost nothing about it. And I suppose one might retort that it serves me right for thinking. All in all, I would sooner be rather ignorant and know it than be twice as ignorant and have a burning zeal for action. I would consider my time well spent if I left you with nothing other than the idea that to overtrump a process that has kept life going, we are assured, for two billion years, using a master plan that we base on a little hasty and ill-informed speculation is, to say the least, a trifle arrogant.

Let us, then, begin at the beginning and proceed slowly and carefully.

1. So far as we know, the major, perhaps the only, permanent impact of genetically deleterious genes is mediated ultimately through reproduction. From an evolutionary standpoint, an abortion, spontaneous or induced; a death in childhood; and a lifelong state, healthy but accompanied by sterility, are all equivalent. Of course, I do not say they are *morally* equivalent; but none of the agitation about screening for Tay-Sachs disease and abortion of homozygotes has any evolutionary impact at all. All it is doing is accelerating a process that, so far as we know, is inevitably going to happen anyway. I have previously called attention to the ambiguity of the Greek word *eugenes,* meaning either "born well" or "well born"—that is, healthy or noble.[9] The term *eugenics* has a precisely similar ambiguity. Early in this century it centered on improving genetic stocks—that is, strategic gain for the whole community. Currently it has come to mean having only healthy children, a tactical gain for the individual family. These are both reasonable objectives, but they are not the same and should not be confused. Parents who have a fetus, homozygous for Tay-Sachs disease, artificially aborted, are doing something for their own benefit and, more indirectly, for that of their immediate family; they are doing nothing for the good of mankind, which must therefore be dismissed as an irrelevant issue.

2. Evolutionary and genetic systems are intensely conservative. In part, this is due to the extremely low rate of mutation, but it is also due in part precisely to the complexity of the system. Even experimental breeders, using methods of selection which are, by natural standards, ruthless, and on characteristics that are reasonably supposed additive, find progress in improving stocks difficult. But where even the polarity of an effect is variable we may expect the results of breeding experiments to be confusing. For instance, one's blood pressure is kept stable from moment to moment by an elaborate set of homeostatic—that is, feedback, or cybernetic—mechanisms. I cannot prove that the existence of these mechanisms is due to genetic factors; but considering their universality, their phylogenetic patterns, and that there are genetic disorders of them known in man, I would

be astonished if they were not heritable. Since they lower the blood pressure if we artificially raise it, and raise the blood pressure if we artificially lower it, it seems quite meaningless to ask whether the mechanism lowers or raises blood pressure. What it does depends on the state of the system. We cannot expect even this device to accommodate itself inside a simple additive system. No doubt but that elsewhere—in the brain, for instance—vastly more complicated systems are at work that must make linearly additive models of intelligence even more precarious. It is not surprising, then, that our puny manipulations of the human population at large should have little strategic impact; and I am disposed to see this conservatism as a special dispensation of Providence to mitigate the presumptuous folly of man.

3. Now why is it all so complicated? Or, to purge my question of teleological overtones, what is the selective advantage of such complexity? So far as we know, the main value is that it produces an ever thicker layer of insulation between the vital processes and the environment and hence enables the organism to survive in a wider and wider range of differing environments. In broad terms, these buffering mechanisms become more elaborate with evolution; and in higher organisms they are more and more supplemented by engineering, including (where man is concerned) air conditioning, the oxygen mask, and passive immunization.

Now the evolutionary consequences of such buffering (predicted theoretically and, in some degree, verified empirically) are interesting. First, and most clearly, all these mechanisms make for more and more multifactorial traits and fewer classical Mendelizing traits. Their capacity for obtunding the force of selection makes for greater tolerance for mutant forms that thus can find more and more hiding places. When every genome contains a good deal of genetic material that is being hidden from the environment, we may find it progressively harder to define what, if anything, we mean by disease; the notion of normality becomes fuzzier and fuzzier.[10] But also, while most mutations—like most new art forms—are deleterious and condemned to ultimate extinction, the protection allows for slow elimination and hence (as it were) extensive evolutionary experimentation. There is every reason to believe that under these protective mechanisms most of the more elaborate evolutionary steps in the higher organisms have occurred. These steps include the development of homeostatic mechanisms themselves, which implies yet more capacity to hide variation and foster evolutionary experiment. Now offhand it is difficult to say whether the gains of complexity, with its protective processes and its opportunities for further development, outweigh the penalties of the amount of junk there must in consequence be in the genome. Boyer[5] seems to have found more variation in the gene pool than can be readily accommodated in classical evolutionary theory, and compilations of variability at biochemical and chromosomal levels are rapidly reaching the level of the individualism encountered in fingerprints.[11] It seems more than likely that evolutionists will soon have to abandon their theory of the payment of debt according to stolid nineteenth-century Republican theories of economics. Indeed, I would venture to guess that homeostasis, like the National Debt or the black hole of astrophysics, is capable of accommodating an indefinite amount of mutational debt. Thus, while not advocating irresponsibility, we should not be unduly swayed by the Jeremiahs who tell us that every child with a genetic disorder that we keep alive artificially is merely another increment to the genetic load, another step to eugenic insolvency. Of course, we must not fall into the logical trap of the a posteriori argument. The fact that so far homeostasis has kept ahead of variation establishes no universal principle of solvency. There is an ascertainment bias. We must not give too much weight to the man who claims that it is quite safe to put his head in the lion's mouth; for if he had been one of the 95 percent to lose their heads, he would not be here to tell us. Dead men tell no tales and write no brochures. The Jeremiahs have a point we should consider—but it is only one point of many.

4. There is another major facet to variation other than as the stuff of evolutionary experimentation. Evolutionary and genetic mechanisms are, as I have said, immensely conservative, because the individual type of mutation is an extraordinary event, and the more precisely we define it, the more extraordinary it gets. At the present level of resolution, a particular mutation occurs once in something like a hundred million births, clearly not a very agile mechanism for responding to an emergency. Now, in the scale of evolutionary time, emergencies are always occurring. An ice age comes and goes in a moment. Smallpox goes in a flash, and malaria, tuberculosis, and leprosy emit at most a brief afterglow. Famine and opulence alternate so quickly as to seem like a vibration. Clearly the cumbersome remedies of evolution cannot respond adequately and promptly to the continually shifting target of excellence. Rather, the collective genome survives by having, like that well-accoutered traveller Colonel Pewter, a little of something for every occasion. It includes fat people who flourish in famines and die of heart attacks in time of plenty. There are profound thinkers who are lionized in sophisticated societies but whose physical ineptness would leave them, in a society of hunters, to be lionized in quite a different sense. There are color-blind people who cannot be deluded by camouflage when hunting but who die driving through red traffic lights in cities, and much more. Just as a society depends economically on a balanced diversity of useful talents, the genome depends for its evolutionary survival on variation, much of which may be despised in one environment but prized in another. One might recall the fable in Aesop of the stag that was proud of its antlers but ashamed of its legs. When the hunt was on, it was its legs that would have

saved it; it was its antlers that became entangled in the branches of the trees.

Of course the discrepancy between new mutation as a source of a suitable genotype and mobilization of genes already hidden somewhere in the bric-a-brac of the genome is, in fact, much greater than I have portrayed. I do not suppose that anybody claims there is one gene for fatness, one for height, and one for resistance to infectious diseases. The relatively speedy adaptation of a species to an environmental change must depend on mobilizing many loci, previously out of work. A whole set of new mutations, each a rare event, and all by chance in the appropriate directions, seems rather too much to hope for in an evolutionary crisis. Thus we conclude that while evolutionary survival demands good averages, it also demands a repertoire of diversity in reserve.

5. I have made no secret of the fact that to manipulate the gene pool is no easy matter. For one thing, its homeostatic structure will tend to offset any such deliberate effect, just as it will counteract random environmental selection. For another, we are woefully ignorant in most cases of how exactly the more complex type of phenotype is produced and hence the most effective way of changing it. For the simplest systems we find, using naive theory, that any impact we have is likely to be slow.[12] Even if we could restore the genetic fitness of, say, phenylketonuria to 100 percent by dietary means, it would take something like 5,000 years to double the frequency of this rare disorder. I cannot undertake to say what will be discovered in the next 5,000 years, but I can think of a great many more urgent problems than what to do with this modest surplus of cases if it is ever attained. In the same way, detecting carriers and telling them not to marry each other will, if successful, have an immediate impact in decreasing affected individuals. However, even if the carriers are totally fit, the day of reckoning for the genetic debt will be millions of generations in the future.

The story with dominant lethals, restored artificially to full fitness, is not quite so innocuous. The frequency will build up perceptibly in the next twenty generations—say, five hundred years—but we do well to remember three things. First, five hundred years is a considerable time; five hundred years in the past separates us from the invention of printing and the heyday of the alchemists. Second, few established dominants are lethal because (by definition) it is very difficult to prove that they are heritable conditions at all. Hence, the impact of improving fitness will be correspondingly blunted. Finally, only a simple and universally effective treatment will produce full fitness. Such ideal treatments, necessarily simple to apply, must be uncommon, and applying them will impose little burden on society where they exist.

By all means let us reflect *soberly* on the consequences of what we are doing in genetic counseling and the prevention of genetic disease. But let us get it into our heads once and for all that we are not talking about an emergency which demands immediate, heroic, and ruthless action. Compared with the problems of pollution, the consequence of genetic manipulation is a nonstarter.

6. So far, I have said nothing at all as to what we should aim for in the gene pool, even if we were in a position to manipulate it appropriately. What are the particular characteristics we should aim for if we do devise the means to pursue them? I have no cogent answer to this problem, merely my set of personal prejudices, what less modest people would perhaps call their principles.[9, 13] I am certainly not going to attempt to deal with this problem here. However, from certain desiderata one may infer certain facts. Thus, if we take survival of the human race as an important value, particular demands seem to follow—for instance, that we must be concerned not only with having good mean values for advantageous characteristics (as the eugenic zealots are) but also with useful variation, that the physical optimum is a perpetually moving target; that one must understand before one acts. Words to be outlawed in such a discussion are *obviously* and *etcetera*. The *obvious* merely gets in the way, as its etymology connotes; *etcetera* merely does what we know we may not do—extrapolate euphorically beyond the limits of the data and the argument. Beyond that, in a spirit of genial democracy, I am prepared to entertain a wide array of opinions so long as proponents are content in the first place to see them as opinions. Formal propositions in this area are usually difficult and often, perhaps, impossible to construct. It is a false inference that the general tone of such propositions is to be gauged from the extremes. Nobody sane wishes to see more babies born anencephalic or acardiac monsters. Nobody with whom we could hold rational discourse will dispute that we wish to foster certain qualities—honesty, courage, compassion, loyalty—provided they are not disproportionately paid for by defects elsewhere. However, we have nothing like the same unanimity over competition, especially among less heroic attributes—for example, whether it is better to be musical or mathematical or whether prowess at football is to be preferred to facility in languages. Indeed, these very disputes and the patronage they engender make for a healthy diversity in society. I think we have had more than enough of the monomania of one central European in this century to have learned a costly lesson on oligarchy of taste; but one generation of public opinion goes nigh to blotting it out, and I hate to think that fifteen million people must be killed every so often to keep this lesson green and lasting in our sad remembrance.

7. Last, we need to consider genetic engineering. What does it have to do with all the foregoing? Surprisingly little. Why, then, discuss it at such

length? Ultimately because, like the Mayor of Hamlin, our losses have made us thrifty. In the long run it is better to avoid, than to repair, costly mistakes. It would be a welcome relief, for once, to be ahead of the reality, to plan how we should use the techniques when they have been perfected, rather than reacting in a blind panic to a crisis announced in a newspaper headline.

Genetic engineering, the way it is currently being discussed, is a brilliant laboratory technique that will, without doubt, prove invaluable in scientific investigation. However, it will not automatically solve any of the problems I have posed—evolutionary, biological, or ethical. It will provide a way of producing certain changes but will not necessarily tell us what the effects would be or what we should try to do.

Again, like all techniques, it deals with the feasible, not the actual. I read with grief some time ago a quotation from a 1920s issue of *Scientific American* about what a wonderful opportunity for education broadcasting would furnish. I wonder if the person who wrote that had any glimpse of radio as a massive force for bad grammar, commercial venality, barbaric taste, and mass-produced prejudices. In principle, the means were available in 1950 to eliminate syphilis entirely, but a quarter century later Frascatoro's hero remains at large. We know much about prevention of cirrhosis of the liver and cancer of the lung, but they remain unprevented. There is no more disappointing facet of human behavior than the perpetual gap between what is attainable, even easily attainable, and what is attained. If logistically simple things fail, we should be even less sanguine about the efficacy of the complicated.

However, one final issue is of current interest, and I have been giving it a little thought. In principle, at least, extracorporeal fertilization, the test-tube baby, offers the opportunity for deliberative and radical eugenics. Prescinding entirely from the issue of whether this procedure is ethical, we perceive that at the stage before fertilization we might pick that gamete from the parent who did not bear the harmful gene. For instance, mothers who are carriers of hemophilia have half their ova bearing the condition and half not. By excluding from fertilization ova with the harmful gene, one is reinforcing the natural phenomenon of prezygotic selection. Or again, if one wished to follow a less disruptive process, one might insert into the genome a normal gene for antihemophilic globulin that would effect a radical cure not only for all the cells in that person's body but also in all descendants from that person. I hope that this kind of argument—I do not for a moment say it is overriding—will be given clear and detailed consideration before we have yet another flood of hasty ethical opinions that linger so embarrassingly on the record when all would wish to forget them.

Conclusions

1. Take your time. There is no eugenic hurry.
2. Who would improve on nature must first understand it. We have a vast amount of homework to do before we are ready to build lofty strategies.
3. Before one can devise the means, one must have first a clear vision of the ends. Action for the sake of action is to be deplored. Humanly speaking, our best appraisal of ends is a consensus among men and women of good will, not a majoritarian view or an esoteric and ex parte decision by any special group.
4. Sufficient unto the day is the evil thereof. We have a far greater responsibility to the actual needs of today than the possible needs of tomorrow. The practicing clinician would do best to reserve his loyalty to his patient whom he has seen rather than to suffering mankind whom he has not seen. The nonclinician would do well to mind his own business. Beware those who can weep for abstractions only!
5. In the long run, through the mercy of Providence, it probably will not make much difference.

Notes

1. E.A. Murphy, ''Genetic and Evolutionary Fitness,'' *American Journal of Medical Genetics* 2 (1978):51–79.

2. V.A. McKusick, *Heritable Disorders of Connective Tissue,* 4th ed. (St. Louis: Mosby, 1972).

3. E.A. Murphy, ''Quantitative Genetics: A Critique'' (Paper presented at the annual meeting of the American Association for the Advancement of Science, Houston, Texas, 3–8 January 1979).

4. S.M. Stanley, ''A Theory of Evolution above the Species Level,'' *Proceedings of the National Academy of Sciences* 72 (1975):646–650.

5. S.H. Boyer, ''Extraordinary Incidence of Electrophoretically Silent Genetic Polymorphisms,'' *Nature* 239 (1972):453–454.

6. K.D. Smith; L. Kunkel; and S.H. Boyer, ''Isolation and Uses of Chromosome-Specific Reiterated DNA,'' *Human Gene Mapping 3* Birth Defects Original Article Series 12 (1976):401–404.

7. O. Kempthorne, ''Introduction,'' in *International Conference on Quantitative Genetics,* edited by E. Pollak; O. Kempthorne; and T.B. Bailey, pp. 3–18 (Ames: Iowa State University Press, 1976); and K. Kidd and K. Gladstein, ''Alternative Genetic Models for the Analysis of Complex Traits'' (Paper presented at Workshop on Facial Clefting, Airlie House, Winchester, Virginia, 4–6 December 1978).

8. R.A. Fisher, "The Correlation between Relatives on the Supposition of Mendelian Inheritance," *Transactions of the Royal Society of Edinburgh* 52 (1918):399; O. Kempthorne, *An Introduction to Genetic Statistics* (New York: John Wiley & Sons, 1957); and D.S. Falconer, *Introduction to Quantitative Genetics* (New York: Ronald, 1960).

9. E.A. Murphy, "The Normal, Eugenics and Racial Survival," *Johns Hopkins Medical Journal* 136 (1975):98–106.

10. E.A. Murphy, "A Scientific Viewpoint on Normalcy," *Perspectives in Biology and Medicine* 9 (1966):333–348; "The Normal," *American Journal of Epidemiology* 98 (1973):403–411; and "On Defining the Normal," in *Strategies in Clinical Research, Mead Johnson Symposium on Perinatal and Developmental Medicine,* no. 9 (1976):7–25.

11. H. Harris; D.A. Hopkinson; and E.B. Robson, "The Incidence of Rare Alleles Determining Electrophoretic Variants; Data on 43 Enzyme Loci in Man," *Annals of Human Genetics* 37 (1974):237–253; and D.S. Borgaonkar, *Chromosomal Variation in Man: A Catalog of Chromosomal Variants and Anomalies* (Baltimore: Johns Hopkins University Press, 1975).

12. E.A. Murphy and G.A. Chase, *Principles of Genetic Counseling* (Chicago: Yearbook Medical Publishers, 1975).

13. E.A. Murphy, "Eugenics: An Ethical Analysis," *Mayo Clinic Proceedings* 53 (1978):655–664.

5 Moral Problems and Ethical Issues in Applied Human Genetics

John C. Fletcher

In the first three sections of this chapter, I describe and review major moral problems in applied human genetics. The activities included here are genetic screening (carrier and newborn), genetic counseling, prenatal diagnosis, and genetic therapies. In the last sections I present some ethical guidance for issues in applied human genetics.

By way of introduction, a distinction needs to be made between the terms *moral* and *ethical*. The following list presents a significant difference between them as the terms represent different dimensions of human experience:

Moral	*Ethical*
What ought to be done in a particular situation,	Reasoning about a moral dilemma that is in the best interest of society and the individuals involved,
A specific and urgent problem,	Generalizability,
Relevant rules for guidance.	Universalizability—reasoning extended to others in similar situations.

A moral problem begins when someone asks, "What ought we to do in this situation?" "Should Mrs. X have amniocentesis?" "Should we treat baby Y for a serious congenital defect?" Moral questions are about concrete and specific cases of conflict, some that are urgent and must be resolved quickly. One of course must perceive or feel that conflict before asking the question. When we are in a moral problem, we typically search for rules or guidelines that are relevant to the features of the problem. For example, if a proper risk to a pregnant woman is involved, we seek guidance about the relationship of risks to benefits.

The term *morality* itself refers to rules or systems of guidance that are supposed to be followed in moral dilemmas—that is, when conflicts of interests, duties, or rules themselves put us in binds that raise questions about how the general welfare will be affected if the dilemma is resolved one way or the other. Moral rules (for example, about fairness, not doing harm, not killing, and so on) are supposed to be followed by everyone without regard to rank, status, or social location. The basic function of morality is to reduce individual expediency as a legitimate reason for action in types of conflicts where the welfare of all is involved in resolving dilemmas. One of the reasons why, in my view, Joseph Fletcher's arguments in situational ethics do not work as well as they might is that he conveys the impression that basic moral rules have less standing in the social process than ought to be the case if every rational person assumed that the rules applied with equal force to self and every other rational person. In his work, Fletcher refers to moral rules and principles as maxims and writes that, if situation ethics has rules, they are only rules of thumb.[1] Although Fletcher stresses that one ought to respect the moral maxims of one's community as illuminators of one's problems, in the absence of a more substantial treatment of the force of morality in society, professional persons especially may take situation ethics to mean that they really did not have to attend to moral rules and that, depending upon the situation, they could do what they wanted to do in the first place. Situation ethics, incorrectly construed, can reinforce an ideology of professionals that they know what is best for the other, or that decision making ought to be done on a case-by-case basis. Although all moral rules have exceptions, this recognition is different when it comes on the other side of a profound respect for the equal force of moral rules on rational persons, rather than centering upon the task of what is meant by being ethical from the perspective of treating moral rules as somehow inferior.

The term *ethical* refers to a different dimension of human experience. Ethics puts the decision maker in the position of being a teacher. He or she should be willing to recommend the reasons upon which the moral dilemma can be resolved to everyone else who will be in a similar situation. To be ethical means to subject one's moral reasons to tests that represent the best interests of individuals involved in such cases and of the community at large. If one is not willing to advocate one's moral reasons in public, especially to others in similar situations, there must be something wrong with the reasons in the first place. Such urgency does not exist in the ethical dimension because, by contrast to morality, a decision does not have to be made that directly affects human welfare. However, a need does exist to get up above (as it were) our decisions and to test their capacity for being generalized to others in similar situations with similar features. The two tests for moral reasons that are most frequently applied in ethics are called the deontological and the consequentialist positions in ethics. More will be said later about these positions and their relevance to moral problems in applied human genetics.

Major Premise

The major premise of my argument is that, in the area of social life that involves human sexuality and reproduction, a prevailing public policy has gradually evolved in societies in the West that is structured by principles of freedom and fairness. The roots of this policy lie in the cultural upheavals of the sixteenth and seventeenth centuries in the West that led to the formation of the modern state and the early shape of the nuclear family.[2] As a result, in this society, human sexuality and reproduction are increasingly identified as voluntaristic activities, especially when viewed against the background of the controlled population sizes of nations in the northern hemisphere. My own ethical views support this policy because it provides common ground to morally pluralistic traditions and minimizes repression of those with dissenting moral views. This public policy is still evolving and open to question, but barring biological or social disasters, it will continue to guide the norms of practice in resolving moral dilemmas in the field of applied human genetics for the foreseeable future.

Moral Problems in Genetic Screening and Counseling

As Dr. Childs and Dr. Hubbard have pointed out in chapters 2 and 3, for centuries the only methods open to humans to control the adverse consequences of natural selection were negative—that is, the avoidance of marriage, reproduction, and intercourse on the basis of social sanctions backed up by verbal pressure. Today physicians and others conduct genetic screening for these purposes:

To detect latent or actual disease prior to making treatment or supportive management available

To detect persons of reproductive age at risk to transmit genetic disease prior to giving information to plan reproduction

To answer research questions about selection, drift, frequency of a genetic marker, natural history of a disorder, etc.[3]

Some of the moral problems that persistently arise in screening are

Problems of veracity in communicating to the community the true purpose of screening. For example, in the early 1970s, many urban communities were told that the purpose of screening for sickle cell trait was to screen 22 million Afro-Americans for a dread disease;[4]

Problems of stigmatization of detected carriers as a result of poor education and information about the meaning of carrier trait;[5]

Problems of lack of informed consent in large-scale screening programs;[6]

Debate about screening affected persons for a disease for which there is no therapy—for example, Huntington's disease;

Problems of confidentiality of findings of genetic screening that result in harm to detected carriers—for example, loss of employment;[7]

Problems in appropriate timing of initiation of screening—for example, some harm was done to children, due to phenylalanine depletion, who were screened and treated by dietary regulation when mandatory screening practices were initiated in the early and mid-1960s.[8]

Figures 5–1 and 5–2 refer us to information about risks of mental retardation related to maternal age and to how a sex-linked disease is transmitted. These figures are relevant to genetic counseling. Genetic counseling should be done only by qualified physicians and genetic associates. The activity involves helping families and individuals at risk for a genetic disorder to understand the medical facts, understand the risk of recurrence, understand the options for dealing with risks, and choose a course of action appropriate to their family goals and values.

Moral problems in genetic counseling converge with those of prenatal diagnosis, but for purposes of identification of these problems, certain ones can be separated.[9] For example:

The issue of paternalism versus autonomy in the process of counseling—for example, how directive or nondirective should the counselor be;

The issue of possible dysgenic (increased frequency of harmful genes) consequences of counseling—for example, two detected carriers decide not to marry each other but marry noncarriers, thus removing the chance that selection would act against the homozygote and thereby increasing the gene frequency;

Conflicts between the welfare of the affected individual and of future generations—for example, if it were possible to identify carriers of the gene for Huntington's disease, these individuals would face the choice of knowingly being diagnosed for a lethal disease of late onset for which no therapy exists as over against choosing not to utilize such a test if it became available;

Problems in truth telling—for example, when family secrets are revealed, such as previous abortions or illegitimacy, informing other fam-

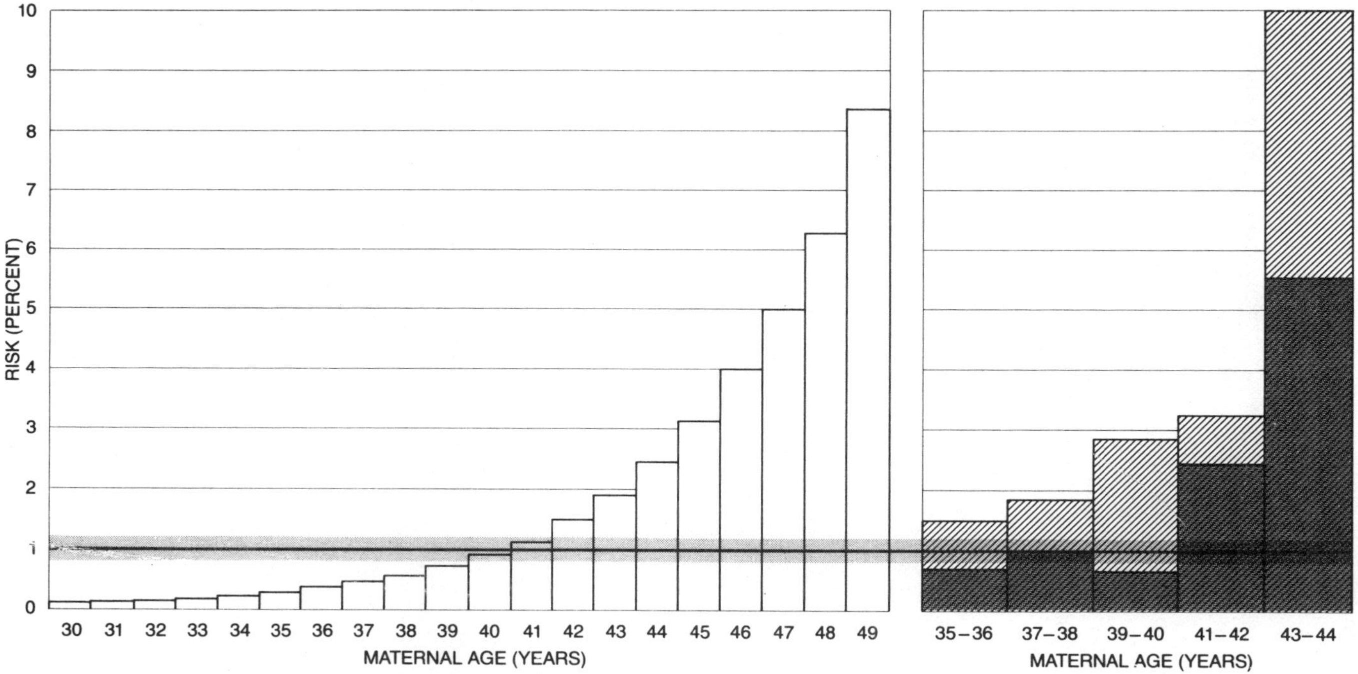

Source: Fritz Fuchs, "Genetic Amniocentesis." 242 (1980):47–53.

Note: Risk of chromosomal abnormality in a fetus is known to increase with the age of the mother. The chart at the left shows the empirical risk that a woman will give birth to an infant whose cells have the most common abnormality: an extra chromosome No. 21. Such infants are born with Down's syndrome. In deciding to do amniocentesis, the risk of such a birth must be weighed against the risk of miscarriage, estimated at 1 percent, that the procedure itself entails. The two risks are in balance if the mother is in her late thirties. If she is forty or older, the risk of miscarriage becomes the lesser of the two. The data are from a study by Ernest B. Hook and Agneta Lindsjö. The bars with hatching at the right show empirical probabilities that any excess of chromosomes (or any deficit) will be discovered by amniocentesis. Darker-gray lower portions on the bars show empirical probabilities that the presence of an extra chromosome No. 21 will be discovered by amniocentesis. Many fetuses with chromosomal abnormalities are miscarried spontaneously. Hence, the bars are higher than those at the left. The data in the chart at the right are from a study by Mitchell S. Golbus and his colleagues.

Figure 5–1. Comparison of the Risk of Fetal Abnormality with the Risk of Miscarriage Associated with Amniocentesis

ily members who may be at risk, and informing the biological parents of an affected adopted child.

A final moral problem on the edge of genetic counseling is posed by whether those who are in a favorable position to recommend counseling ought to do so. Health professionals in the workplace and schools, clergy who prepare couples for marriage, public-health nurses, and physicians who meet patients at the first obstetrical interview are examples of well-positioned professionals. Clergy officiate at 80 percent of all first marriages in the United States and 60 percent of all second marriages.[10] They, like other professionals, can choose not to make persons who call on them for help aware of the benefits of genetic counseling. However, there is a moral problem in not doing so. They are in a position to refer persons to help that may result in the prevention of great harms as well as reduce the chances that spouses will blame one another for birth defects in children that might have been avoided had they known more prior to marriage. Most clergy probably do not feel professionally prepared to raise such an issue, or their teachers have not encouraged them or instructed them on how to refer couples to genetic counselors. The moral issue is clear when a clergyperson has the occasion to ask, following the birth of an affected child to a couple who prepared for marriage with him or her, "Ought I to have raised the question of genetic counseling with this couple?" The reason for bringing up the subject, knowledge that could possibly prevent harm as well as strengthen their relationship, far outweighs (in my view) reasons that might be used against it.

The information in figure 5–2, about the mechanisms by which sex-linked genetic disease (for example, Lesch-Nyhan, hemophilia, Duchenne's muscular dystrophy) is transmitted, brings up another example of a moral problem that recently came to my attention—namely, a counselor who told a so-called therapeutic lie in order to relieve a mother of the guilt of having transmitted the disease. This counselor invented a story that the father contributed freshly mutated genes, as a partial cause, to a situation of X-linked disease. The counselor was omnipotently, in my view, attempting to assume responsibility for the well-being of this family and the supposed marital happiness that ought to ensue from such an act of deception. The pretentiousness of this act is exceeded only by the ignorance of the fact that such couples do much reading and research on their problems. Of course, they can read and will undoubtedly discover the contradiction between what they are told and what they read. Counselees should not be so patronized, and there are definite limits to the counseling role, which does not include such therapy for the marital relationship.

Ethical Issues in Prenatal Diagnosis

Figure 5–3 illustrates the major method of prenatal diagnosis—amniocentesis, a needle puncture of the amniotic sac—to draw off a small amount (25–30 ml) of amniotic fluid from which fetal cells can be cultured for study. More than 100,000 amniocenteses have been done to date throughout the world. Other methods of diagnosis include ultrasonographic fetal imaging and fetoscopy. The latter technique is used to retrieve fetal blood to diagnose diseases of the hemoglobin such as sickle cell disease.

The ethical issues that have most frequently arisen in prenatal diagnosis are reviewed in the following list:

Selective abortion:
- Justification,
- Election,
- Link to discontinuation of neonatal therapy;

Risks:
- Safety, accuracy of techniques,
- Accurate communication, consent,
- Adequacy of counseling and psychiatric support pre-/postabortion;

Controversial indications:
- Appropriateness for individuals,
- Anxiety,
- Sex selection;

Fairness of access.

Selective Abortion

Although parents and physicians receive good news in more than 95 percent of prenatal studies in the sense that the disease for which the fetus may be at risk is ruled out, the remainder receive bad news. The positive diagnosis does not determine the options, one must remember, but rather the options are structured by the beliefs and values of the parents. Abortion is one option. Others are to prepare for intrauterine treatment of the fetus in a very few instances, or to prepare for delivery with or without treatment and with or without plans for adoption.

The moral dilemma about abortion affects both parents and physicians. Parents are caught between the rightness of being concerned about the

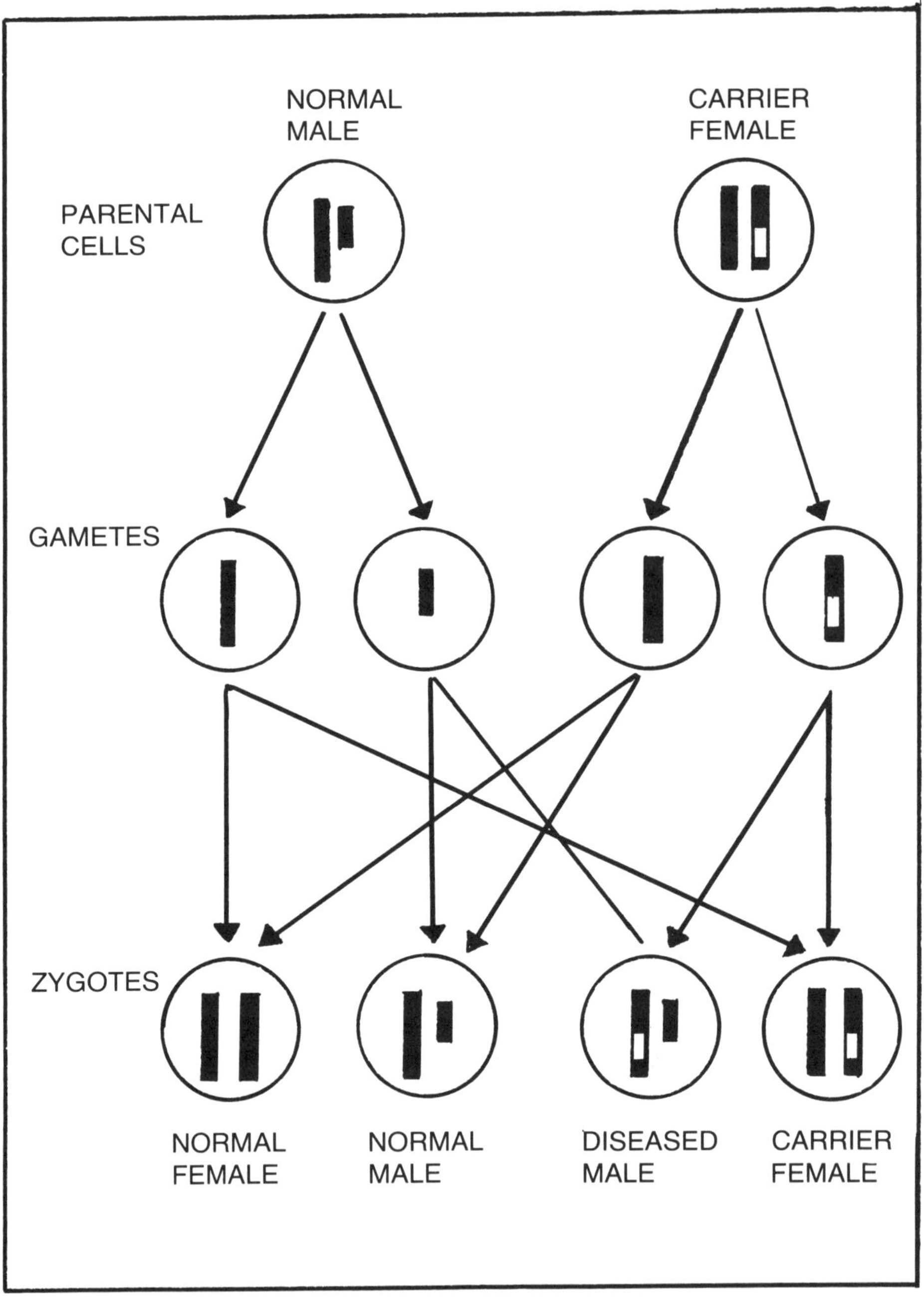

Note: The light bar on the X chromosome indicates the presence of a malfunctioning gene.

Figure 5–2. Pattern of Inheritance for Sex-Linked Disorders

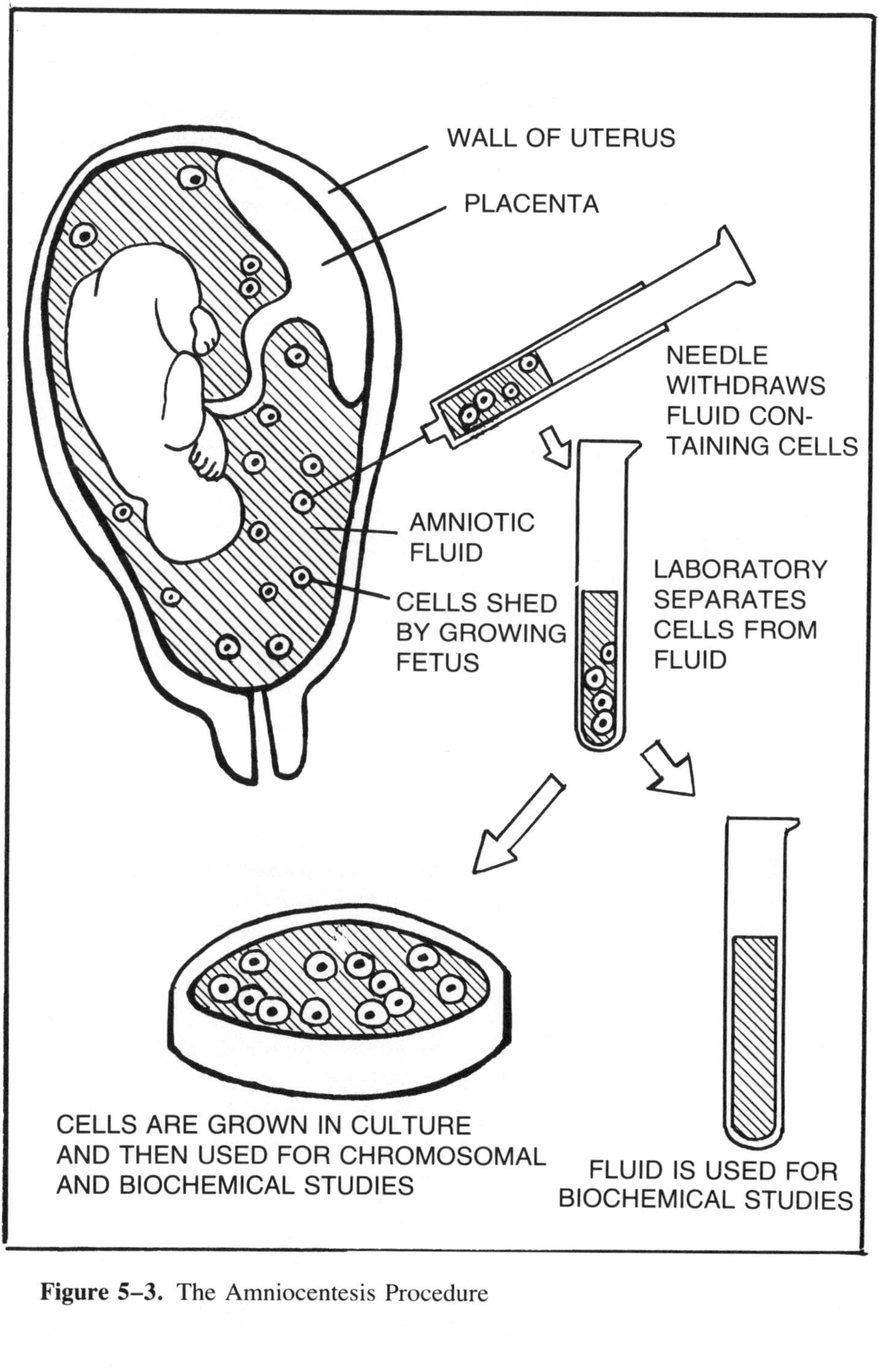

Figure 5–3. The Amniocentesis Procedure

welfare of their wanted child and the rightness of protecting themselves and their families against disaster. If the diagnosis is positive, they cannot obtain both good outcomes. Physicians are taught that they ought to want to cure disease. At the same time, they are taught to prevent disease if possible. Ethicists and physicians who morally oppose prenatal diagnosis do not oppose the technique as such but the use to which technology extends (abortion), which they hold is incompatible with the purpose of saving life that constitutes medicine.[11] Those who support selective abortion base their support on the obligation to prevent suffering and disease.[12] These two contending affirmations, however, reveal the two sides of the moral conflict.

Risks

The direct risks of prenatal diagnosis by amniocentesis—fetal death or infection in the mother—were sources of moral concern in the early use of the technique. Widely accepted controlled studies showed that fetal death is less than 1 percent from the procedure in the United States, Canada, and the United Kingdom.[13] These risks are higher for fetoscopy, but results improve greatly with experience.[14] False positive and false negative rates due to human error and laboratory contamination are small, but real, risks. The moral content of the risk issue arises from the duty to avoid harm and to increase benefits.

A second group of risks, indirect and secondary, has also been identified with prenatal diagnosis by ethicists who link decision making about the affected fetus with the high-risk newborn with the same degree of handicap.[15] Is there a morally relevant difference between the fetus with a diagnosed open-neural-tube defect and a newborn with the same impairment? Does permitting abortion of the fetus warrant pediatric euthanasia? The opposite question is: If one would not permit active euthanasia with the newborn, how can one morally abort a fetus at an earlier stage for reason of disease?

I have argued in the past that one morally relevant difference exists between the fetus and the newborn—namely, that the latter is separate from the mother.[16] The separateness criterion provides physicians and others with a reason to protect the life of the newborn even when parents disagree with a medical strategy because the mother's best interests can be considered independently of the child and vice versa. During pregnancy, the two sets of interests are not truly separable. The problem with this argument is that advancing technology may render it moot. Newly fertilized embryos are separate before reimplantation. Questionably viable premature infants with extremely low birth weight are separate when supported technologically. The concept of viability itself has been rendered less applicable to neonatal decision making because of the extreme dependence of the high-risk newborn on mechanical means of support and long periods of hospitalization. We are also entering a period when the fetus with a congenital defect will increasingly be regarded as a patient who can be treated in some cases.[17] Because it is ethically undesirable for technological change to guide and control norms for moral choice, arguments from viability and separateness are weakened. Possibly a stronger case can be made to protect the rights and the life of the fetus in late gestation and after delivery on the basis of avoiding the various harms of pediatric euthanasia rather than on the basis of moral differences in the status of the fetus and the newborn, especially when this moral status is subject to so many considerations introduced by technology. A more extensive discussion is appropriate at another place and time, but the reader should know that I am less persuaded by the force of arguments based on differences in the moral status of the fetus and the newborn.

Another risk of abortion following prenatal diagnosis is the consequent depression in both parents, documented by Blumberg and others, which requires psychiatric intervention to alleviate.[18] This reaction is strong evidence that such parents desire children deeply. A prenatal-diagnosis center should be prepared to offer help for such parents in a timely way.

Controversial Indications

Two controversial requests for amniocentesis that fall outside the usual medical indications give rise to a moral problem. The first is when the mother is simply anxious about the possibility of mental retardation and when no family history supports the request. A second is for sex selection unrelated to sex-linked disease. Reasons used by physicians and others to oppose granting such requests include sex discrimination, wasting scarce resources, trivialization of serious reasons required for abortion, and misuse of a medical tool.[19] Why does a moral problem exist here at all? There is a moral problem if the physician does not want to single out abortion for sex choice as more blameworthy than abortion for social or economic considerations. In order to prevent sex-choice abortions, legal or administrative steps would have to be taken that, if enacted, would doubtless be precedents for further restrictions. If one is interested in protecting the existing social policy on abortion, one would not absolutely oppose such requests but rather construct procedures to discourage them while remaining open to those that may be justified on psychiatric grounds.

Access

In my estimation, the most pressing ethical and moral issues of the 1980s with respect to prenatal diagnosis are as follows:

Growth and demand;

Socioeconomic priority of genetic services;

Gene therapies:
Direct,
Enzyme replacement.

Demand for prenatal diagnosis will grow because the population itself is aging and maternal/paternal age is relevant to risk of retardation. Yet studies of the use of prenatal diagnosis show that many more highly educated and affluent users exist than their opposites.[20] Costs of services also pose a significant barrier to less-affluent or noninsured users. Legal barriers exist to the use of Medicaid funds to reimburse for abortion for genetic disease. The moral problem here is the offense to distributive justice in that the benefits of applied human genetics are unevenly distributed so that the burdens of genetic disease fall disproportionately on poor and ill-educated individuals. Also, Medicaid funds are used to reimburse for treatment of conditions that cause far less damage to an infant and family than many genetic diseases that can be diagnosed. Medicaid payment for prenatal diagnosis is practically nonexistent, and third-party payment for the procedure varies widely.

Access to the benefits of screening pregnancies for open-neural-tube defects by studies of maternal-serum alpha-fetoprotein has been delayed for a long period due to disagreement about whether the medical community can provide and coordinate the collateral services needed to optimize such tests.[21] Reports from the United Kingdom and pilot programs in the United States show that this form of testing is safe and effective.[22]

The question of access to human genetic services also raises the issue of what priority the society and its policymakers want to place on this type of medical intervention. In the light of competing goals of health care and the fact that the nation has not yet achieved an assurance of minimal health care for each citizen, it is not likely that the priority for genetic services will be high in the near future. Also, a higher priority for genetic services is linked to the possibility of therapy for genetic disease, an alternative now real only in a very few instances.

Ethical Guidance

A body of literature on the moral problems of applied human genetics, providing ethical guidance to those who are in leadership positions and those who are in need of the knowledge and techniques of practitioners in the field, gradually accumulated in the 1970s. The literature divides mainly into two types: (1) articles and books by individual geneticists or ethicists that contribute to the discussion and debate about moral problems and reflect the author's own ethical approach to the problems and (2) reports from task forces assembled by research centers in bioethics, scientific, and religious bodies for the review and evaluation of the moral problems of applied human genetics. The reader can find perhaps the best review of the first type in the *Encyclopedia of Bioethics*.[23]

The earliest statement of ethical guidance for genetic screening was issued by a research group of the Hastings Center in 1973 that reflected an emergency social situation created by hastily fashioned sickle cell trait and disease screening.[24] The major ethical themes in the document are the autonomy of the individuals being screened, the voluntariness of the programs, and the protection of individuals from potential harm that might ensue from a lack of confidentiality. The public-health goal of screening (prevention of disease) was subordinated to the benefits to individuals and families that could come from screening. The autonomy of individuals and voluntary participation was to be protected by informed consent to the extent that the authors of the statement believed genetic screening to be a form of human experimentation and thus to be guided by the consent requirements of federal practices. The freedom of physicians and community leaders to begin screening at all was not challenged on ethical grounds, although the statement referred to the unresolved abortion controversy.

In 1975, a report from the National Academy of Sciences brought together the best scientific, medical, and historical reflection on screening at the time.[25] Ethical guidance stressed the exclusion of political and eugenic goals and an insistence on informed consent of those being screened. The same themes that appeared in the Hastings guidelines reappear here. In 1977, Holtzman's study of newborn screening for genetic-metabolic diseases appeared under federal auspices.[26] The report contained a short statement on the ethics of screening and how to obtain informed consent from parents.

No definitive set of guidelines for genetic counseling exists that is comparable to the Hastings screening statement and a later statement on prenatal diagnosis. An extensive volume on the moral, legal, and social issues in genetic counseling was published in 1979 by the National Foundation-March of Dimes under the auspices of the Hastings Center.[27] Lebel discussed extensively the ethical issues in the genetic-counseling relationship.[28] The ethical guidance given in both of these discussions stresses the freedom of the counselee and the restraint of the counselor in their interactions. A definition of genetic counseling adopted by the American Society of Human Genetics accentuated the autonomy of the family members, in the light of their ethical and religious standards, to choose the course of action that "seems appropriate to them in view of their risk."[29]

As in the case of genetic screening, the Hastings Center was the first to organize a task force to formulate guidelines for the conduct of prenatal diagnosis.[30] The report defines prenatal diagnosis as one type of screening and offers criteria that correlate strongly to the earlier screening guidelines. The task force clearly favored use of prenatal diagnosis for certain "well-defined groups of pregnant women known to be at risk." On the crucial issue of abortion, the document reflects "a moral framework favoring the protection of individual choice and the autonomy of parents, even when we disagree with their courses of action." Following this affirmation of the principle of freedom, the diversity of views on abortion is reviewed along with an intent to respect the pluralism of values on the subject. The theme of fairness is raised as a standard for the conduct of prenatal diagnosis and illustrated by case examples—for example, prenatal diagnosis should not be denied to a woman at risk who has decided against abortion at the outset.

The Hastings guidelines on prenatal diagnosis are mentioned positively in the report of a Consensus Development Conference of the National Institutes of Health held on the subject of antenatal diagnosis in 1979.[31] The official summary of this conference emphasized that the decision about the procedure should be made by the family. A report issued by the National Council of Churches about applied human genetics tries to balance the interests of the family or individual and the interests of society as it evaluates the moral problems of prenatal diagnosis.[32] Rather than giving specific guidelines, the report encourages an approach to ethical decisions that would help prevent suffering but not at the cost of making suffering an "enemy of human dignity." The report of an international workshop on prenatal diagnosis convened in Canada in 1979 recommended that the Hastings Center guidelines be applied in any new prenatal-diagnosis programs.[33] The final form of ethical guidance that appeared in the last decade was guidelines for prenatal diagnosis involving members of minority groups in the United States.[34] This statement, also done by the Hastings Center, extended the principles of the earlier statement to the protection of the freedom of minority-group members, including favoring parental choice about abortion following a positive finding.

Conclusion: Ethical Guidance, Present and Future

The previous section shows that bodies of ethical guidance were developed in the 1970s for applied human genetics that depend mainly upon the ethical principles of freedom and fairness. Freedom is understood both positively and negatively. On one side, freedom means the inhibition of external restriction, harm, and the disabling of voluntaristic activity. On the other side, freedom means self-realization and the satisfaction of basic needs. The autonomy of the individual and its protection involves both sides of the freedom principle. Fairness is understood in this context as impartial and equal treatment of those who participate in voluntary activities, as well as the source of the obligation to measure up to the standards expected of all who participate in the activity. Freedom and fairness are the ethical principles that form the basis for the social policy that presently functions to help resolve the moral problems in applied human genetics. For example, affected individuals are neither coerced into screening nor forced to submit to genetic advisors. Persons at risk to transmit genetic disease may marry others without legal sanctions. Neither abortion nor sterilization is forced upon individuals; on the contrary, these are entirely dependent upon the voluntary request of the individual.

Ethical guidance based upon freedom and fairness are not the only statements that exist in the public forum. A countervailing challenge to the existing social policy is the contemporary effort for a constitutional amendment limiting freedom in abortion decisions. McCormick, one of the most respected critics of the prevailing policy as "excessive individualism and autonomy," outlined a proposal for an amendment based upon the difference in abortion if "the alternative is tragedy" or abortion for "mere inconvenience."[35] Although there is a ring of truth in the claim that some reasons for abortion are more weighty than others in terms of a justification for an exception to the general rule against killing, one must object to the proposal on the grounds of the greater damage that would be done to the cause of fairness when others than a woman and her physician would make judgment that the case was tragic and not inconvenient. Each abortion request would need to be adjudicated. If such were the case with abortion, then why not other decisions having to do with reproduction, such as sterilization and artificial insemination by donor? To make a case for a change in the public policy on abortion, one needs to show that the damage being done to the society and its members by the policy outweighs the damage that would be done if the policy were reversed. This case has not been made and cannot be made persuasively as long as the principles of freedom and fairness continue to inform and structure concepts of what are the best interests of individuals and society in reproduction or human sexuality.

The historical roots of this public policy lie in changes in marriage and family relations in European nations of the sixteenth and seventeenth centuries following the Protestant Reformation.[36] These movements also coincided with the rise of the modern state. Gradually, the freedom of the individual to choose his or her own marriage partner, apart from paternalistic coercion, was increasingly enjoyed by more members of society. The principle of fairness grew increasingly relevant to resolving conflicts between men and women, as well as between children and their parents. Since this period, sexuality and reproduction have increasingly been defined as voluntaristic activities. In the background of these developments in culture is the fact that the populations of Western nations have, demographically considered, become more controlled with respect to fertility. Without this background of controlled fertility, it is doubtful if the principles of freedom and fairness would have wide acceptance in the population or by policy-makers. More coercive or stringent methods would obtain. Ethics and values interact with a biological and cultural substrate in an evolutionary pattern.

As technological and medical controls on reproduction emerged in their twentieth-century forms, these means were adapted within the context of the social norms guiding sexuality and reproduction. Knowledge and technology has been diffused through a public filter of individual choice. My argument is that such a policy is the best available public policy because it affords possibilities for the coexistence of rival ethical views with the least amount of public punishment or repression for any persons whose views differ from the prevailing view. The dynamics of public participation in the observance of the policy are quite similar to those in religious toleration. An axiom that is similar to one that guides religious toleration can be thus phrased: I will respect your freedom to choose to avoid or achieve reproduction with or without technological intervention, if you will respect my freedom to do otherwise. Further, I will work for a society in which neither of us will suffer harm or punishment because of the differences that exist when freedom is cherished.

This axiom attempts to embody elements from the two major ethical positions that actually function in our society to guide resolutions of moral problems. The deontological position is based on the concept that certain actions are intrinsically good or evil. The term itself is derived from the Greek word for "duty." Our primary duty, in this view, is to avoid evil acts and do good ones. The consequences that flow from doing or avoiding are to be seen as secondary to the actions themselves. It is wrong to do an abortion, in this view, regardless of the consequences. The second position is usually called consequentialism. Our primary obligation, in this view, is to promote the best set of consequences possible in each situation. If widely accepted moral rules or guidelines would prevent the best set of consequences, the consequentialist would recommend acting for the good in spite of the rules. For instance, if lying to one person would prevent predictable harm to another, it would clearly be worth breaking the rule against lying. The primary ethical test for a deontologist's decisions is: Can I recommend my actions themselves to others who are in similar situations as actions they should adopt under any and all circumstances? The primary ethical test for a consequentialist's decisions is: Can I show that the consequences I will achieve by my actions in this situation are so demonstrably good that the actions I took to achieve the good results will easily be justifiable to others in similar situations?

Clearly, if these two positions are set up as absolute opposites, there is no room for reconciliation. In public debate they are often presented in just this way. However, in practice and in policymaking, elements from both positions are incorporated into solutions. In the field of research with human beings, the consequentialist view is strongly represented by the risk/benefit assessments done by institutional review boards; but the deontological position is also strongly represented by the obligation to obtain the voluntary informed consent of the subject. No person ought to be coerced into research, regardless of the good that might be done by the research.

In just this way, elements from both positions are incorporated in the social policy of freedom with fairness that presently guides choices in the context of applied human genetics. The deontological position is represented by the lack of coercion of individuals to be screened or counseled, regardless of the consequences of being constrained to intervene in the life of an affected individual. A couple at risk for recessive disease can marry and reproduce, without being punished, if they object to screening and counseling. In spite of the benefits that could be achieved by more aggressive public-health measures, under present circumstances it would be considered unethical to require all women over forty who become pregnant to have prenatal diagnosis. The damage that would be done to individual rights and privacy by such an intervention is considered by most to be more serious than the prevention of mental retardation that would be achieved by the screening.

The consequentialist position is represented in the prevailing policy because of the underlying assumption that the technology itself is not intrinsically harmful but rather a means to an end of preventing calculable suffering. Without the techniques and methods of applied human genetics, the total burden of suffering in the society would be significantly increased. For those at-risk families who choose to use genetic services and employ abortion to prevent recurrence of disease, decisions are validated on the basis of enlarging their options and protecting the family from harm.

In my own view of ethics, resources from both positions are required to guide the complexities of the interactions of society and science. I would be the first to acknowledge that the policy of freedom with fairness exists in the context of an affluent and technologically strong society with a population that is in a relatively controlled position when compared to economies and populations in the southern-hemisphere nations. The greatest threat to the existing policy is economic or social disasters that might turn the strong against the weak. Although the populations of Western nations are controlled, their economies are strong enough at present to support the risk inherent in refusing to coerce parents who are genetically at risk to refrain from reproduction. Given disastrous economic scenarios, it is not beyond imagination that rationing of all types would include limiting the freedom to replicate a life-threatening and expensive disease.

The second threat is abolition of the freedom of choice about abortion. If abortion for genetic defects were illegal, the only option open to parents of an affected fetus when no therapy exists would be to prepare for birth. Society would coerce women with seriously compromised pregnancies to go to term. If a future society were willing to use this much coercion in this instance, how much more would be used in other areas of social and interpersonal life?

The great virtue of the existing policy is that it allows common ground for those who are on opposite sides of the ethical debate to choose to use or not to use the techniques of applied human genetics without punishment or harm for either choice. An ethical absolutism on either side of the two ethical positions that inform the policy will cause this common ground to disappear.

Notes

1. J. Fletcher, *Situation Ethics* (Philadelphia: Westminster Press, 1966), pp. 26, 36.

2. L. Stone, "The Rise of the Nuclear Family in Early Modern England: The Patriarchal Stage," in *The Family in History*, edited by C.E. Rosenberg, pp. 12–57. (Philadelphia: University of Pennsylvania Press, 1975).

3. National Research Council, *Genetic Screening, Programs, Principles, and Research* (Washington, D.C.: National Academy of Sciences, 1975).

4. R.M.Nalbaldian, "Mass Screening Programs for Sickle Cell Hemoglobin," *Journal of the American Medical Association* 221 (1972): 500.

5. G. Stamatoyannopoulos, "Problems of Screening and Counseling in the Hemoglobinopathies," in *Fourth International Conference on Birth Defects*, edited by A.G. Motulsky and F.J.B. Ebling, International Congress Series No. 297 (Amsterdam: Excerpta Medica Foundation, 1973).

6. J.C. Fletcher; R.O. Roblin; and T.M. Powledge, "Informed Consent in Genetic Screening," in *Birth Defects: Original Article Series, Vol. 10, Ethical, Social and Legal Dimensions of Screening for Human Genetic Disease,* edited by D. Bergsma (1974), p. 145.

7. C.F. Whitten, "Sickle Cell Programming—An Imperiled Promise," *New England Journal of Medicine* 288 (1973):318.

8. A National Academy of Sciences Committee for the Study of Inborn Errors of Metabolism concluded (see note 3) that "hindsight reveals that screening programs for phenylketonuria were instituted before the validity and effectiveness of all aspects of treatment, including appropriate dietary treatment, were thoroughly tested" (p. 2).

9. F.C. Fraser, "Introduction: The Development of Genetic Counseling," in *Birth Defects: Original Article Series, Vol 10, Genetic Counseling: Facts, Values and Norms,* edited by A.M. Capron, National Foundation-March of Dimes (1979), p. 10ff.

10. National Center for Health Statistics, Department of Health, Education, and Welfare. *First Marriages. United States 1968–1976,* DHEW Publication No. (PSH) 79-1913, September, 1979, p. 5.

11. P. Ramsey, "Screening: An Ethicist's View," in *Ethical Issues in Human Genetics: Genetic Counseling and the Use of Genetic Knowledge,* edited by B. Hilton, pp. 147–161, Fogarty International Proceedings no. 13. (New York: Plenum Press, 1973); and J. Lejeune, "On the Nature of Man," *American Journal of Human Genetics* 22 (1970):121–128.

12. J. Fletcher, *The Ethics of Genetic Control* (Garden City, N.Y.: Doubleday, Anchor Press, 1974); and A.G. Motulsky, "The Significance of Genetic Disease," in *Ethical Issues in Human Genetics,*edited by B. Hilton, pp. 59–66 (New York: Plenum Press, 1973).

13. NICHD National Registry for Amniocentesis Study Group, "Midtrimester Amniocentesis for Prenatal Diagnosis: Safety and Accuracy" *Journal of the American Medical Association* 236 (1976):1471–1476; N.E. Simpson; L. Dallaire; J.R. Miller; L. Siminovich; J.L. Hamerton; and C. McKeen, "Prenatal Diagnosis of Genetic Disease in Canada: Report of a Collaborative Group," *Canadian Medical Association Journal* 115 (1976):739–748; and Report to the Medical Research Council by their Working Party on Amniocentesis, "An Assessment of the Hazards of Amniocentesis," *British Journal of Obstetrics and Gynecology* 85 (1978):Supp 2.

14. U.S Department of Health and Human Services, National Institutes of Health, *Prenatal Approaches to the Diagnosis of Fetal Hemoglobinopathies,* NIH Publication No. 80-1529, May 1980, p. 13ff.

15. P. Ramsey, "Reference Points in Deciding about Abortion," in *The Morality of Abortion,* edited by J.T. Noonan, pp. 60–100 (Cambridge, Mass.: Harvard University Press, 1970).

16. J.C. Fletcher, "Prenatal Diagnosis, Selective Abortion, and the Ethics of Withholding Treatment from the Defective Newborn," in *Genetic Counseling: Facts, Values, and Norms,* edited by A. Capron, p. 250 (New York: Alan P. Liss, 1979).

17. M.S. Golbus, *Prenatal Diagnosis* (White Plains, N.Y.: March of Dimes—Birth Defects Foundation, 1980).

18. B.D. Blumberg; M.S. Golbus; and K.H. Hanson, ''The Psychological Sequelae of Abortion Performed for a Genetic Indication,'' *American Journal of Obstetrics and Gynecology* 122 (1975):799–808.

19. A. Milunsky, *Know Your Genes* (Boston: Houghton Mifflin Company, 1977), p. 277.

20. D.C. Sokal; J.R. Byrd; A.T.L. Chen; M.F. Goldberg; G.P. Oakley, ''Prenatal Chromosomal Diagnosis: Racial and Geographic Variation for Older Women in Georgia,'' *Journal of the American Medical Association* 244 (1980):1355.

21. National Center for Health Care Technology, U.S. Department of Health and Human Services, *Maternal Serum Alpha-Fetoprotein. Issues in the Prenatal Screening and Diagnosis of Neural Tube Defects* (Washington, D.C.: Superintendent of Documents, U.S. Government Printing Office, 1980), p. 84 f.

22. Working Group on Screening for Neural Tube Defects, *Report* (London: Department of Health and Social Security, 1979); and J.E. Haddow and J.W. Macri, *Screening for Neural Tube Defects in the United States* (Portland Me.: Pilot Press, 1977).

23. W. Reich, ed. *Encyclopedia of Bioethics,* 4 vols. (New York: Macmillan and Free Press, 1978).

24. Research Group on Ethical, Social, and Legal Issues in Genetic Counselling and Genetic Engineering, ''Ethical and Social Issues in Screening for Genetic Disease,'' *New England Journal of Medicine* 286 (1972):1129.

25. National Research Council, *Genetic Screening.*

26. N.A. Holtzman, *Newborn Screening for Genetic-Metabolic Diseases,* DHEW Publication No. (HSA) 77-5207, 1977.

27. F.C. Fraser, ''Development of Genetic Counseling.''

28. R.R. Lebel, ''Ethical Issues Arising in the Genetic Counseling Relationship,'' *Birth Defects, Original Article Series,* 15, (1978).

29. Ad Hoc Committee on Genetic Counseling, ''Genetic Counseling,'' *American Journal of Human Genetics* 27 (1975):240–242.

30. T.M. Powledge and J.C. Fletcher, ''Guidelines for the Ethical, Social and Legal Issues in Prenatal Diagnosis, *New England Journal of Medicine* 300 (1979):168.

31. Consensus Development Conference, National Institute of Child Health and Human Development, Antenatal Diagnosis, NIH Publication No. 79-1973, 1979, pp. 1–196.

32. National Council of the Churches of Christ in the USA, *Human Life and the New Genetics* (New York, 475 Riverside Drive, 10027, 1980).

33. J.L. Hamerton and N.E. Simpson, *Prenatal Diagnosis, Past, Pres-*

ent and Future, in *Prenatal Diagnosis (Special Issue) Report of an International Workshop* (New York: John Wiley & Sons, 1980), p. 56.

34. R.F. Murray; N. Chamberlain; J. Fletcher; E. Hopkins; R. Jackson; P.A. King; and T.M. Powledge, ''Special Considerations for Minority Participation in Prenatal Diagnosis,'' *Journal of the American Medical Association* 243 (1980):1254.

35. R.A. McCormick, ''Abortion: A Changing Morality and Policy,'' *Hospital Progress* 60 (1980):36–44.

36. L. Stone, ''Rise of the Nuclear Family,'' pp. 12–57.

6 In Vitro Fertilization and Embryo Transfer in Human Beings

John D. Biggers

A girl was born on July 25, 1978, as a result of in vitro fertilization and embryo transfer.[1] After that, a boy was born in England through the efforts of the same investigators,[2] and more recently a girl was born in Australia through the efforts of an independent team. Another girl may also have been born in India. The birth of the first of these babies caused worldwide publicity that led simultaneously to acclaim and concern. The acclaim greeted the development of another technique to treat barrenness due to blocked fallopian tubes—one of the common and intractable causes of human infertility. The concern arose over ethical questions involving the possible creation and destruction of human beings, the normality of the offspring, a change in the natural patterns of human reproduction, and the possibility of genetic engineering in our species. This chapter attempts to evaluate the present status of the procedure from a biomedical point of view, to provide a proper understanding of the potential usefulness of the method, and to create a sound scientific basis for the discussion of the moral issues that are frequently raised.

Fertilization takes place in the upper third of the oviduct, the ampullary region. The early embryo then commences a series of mitotic divisions and cleaves into a morula. Later, fluid accumulates within this ball of cells, and the embryo becomes a blastocyst. Throughout this period, the embryo floats freely in the fluids in the female tract, traveling down the oviduct and finally entering the uterus. This period is called the preimplantation period of pregnancy. It is terminated by the implantation phase, in which the blastocyst adheres to the endometrium and begins to invade the maternal tisses. From the 15 human embryos that have been recovered before implantation, it seems that the embryo takes three days to traverse the oviduct. By this time the embryo has developed to the 12-to-16-cell stage and is known as an early morula.[3] Implantation occurs six days after fertilization, so that the human embryo is unattached in the uterine secretions for three days while it develops to the blastocyst stage.

Infertility caused by disease of the oviducts has been recognized for a long time. Several approaches have been introduced over the years to treat

---

Reprinted by permission of the *New England Journal of Medicine* 304 (1981):336–342.

women with normal ovaries and a normal uterus, but bilaterally occluded or absent oviducts. Mackenrodt is reported to have treated two cases successfully by salpingostomy in 1884,[4] and Morris[5] grafted the ovary into the uterus in 1895. The latter procedure was the forerunner of Estes's operation.[3, 6] Neither of these procedures proved very effective; thus, infertility specialists have always been receptive to alternative treatments. Nowadays, major interest is being focused on the repair of oviducts by microsurgery and on the possible use of extracorporeal fertilization.

The idea of extracorporeal fertilization with gametes from the infertile couple and replacement of the embryo in the normal uterus of the woman was suggested 43 years ago in an unsigned editorial in the *Journal*.[7] However, at that time the basic research that was necessary to make the proposal scientifically viable was in its infancy. Proof that fertilization of mammalian ova could occur in vitro was lacking, and no methods for culturing the mammalian ovum through the cleavage stages had been developed. The several claims of in vitro fertilization made in the 60 years since the discovery of fertilization were equivocal. For example, light microscopy could not provide definitive evidence of penetration of an ovum by a sperm.[8] The only important contribution had been made many years earlier in 1890 by Heape,[9] who showed that early rabbit embryos could be flushed from the oviduct with physiologic saline and transferred to a foster mother, where they would resume normal development and be born. By 1959, the necessary basic research had been done. Hammond[10] showed that it was possible to culture early mouse embryos from the eight-cell stage to the blastocyst in a complex biologic medium. A few years later, Whitten demonstrated[11,12] that mouse embryos could develop from the two-cell stage to the blastocyst in a simple medium. The following year, McLaren and Biggers[13] showed that mouse blastocysts produced by Whitten's procedure would develop into normal, fertile adult mice after transfer into the uteri of foster mothers. Although many claims of successful fertilization of mammalian eggs in vitro had been made over the years, it was not until 1959 that this event was unequivocally proved by Chang in rabbits by means of genetic markers.[14]

Edwards began his work on the human egg in about 1963,[2] and his first paper on the subject was published in 1965.[15] He has collaborated with Steptoe since about 1968,[2] and their first joint paper was published in 1970.[16] Their persistence over a decade and a half resulted in the birth of Louise Brown, thus realizing the predictions made in 1937.

Overall Results of Steptoe and Edwards

The procedures developed by Steptoe and Edwards consist of three steps: recovery of an ovum that has completed meiotic maturation, fertilization of the mature ovum in vitro and subsequent culture of the embryo through part of its preimplantation development, and replacement of the embryo in the uterus of the patient from whom the ovum was obtained. This procedure differs from that usually performed in laboratory and domestic animals in that the embryo is placed in the uterus of the patient, rather than in the uterus of a surrogate mother.

The extensive clinical work of Steptoe and Edwards falls into two phases that are related to the methods adopted for the recovery of meiotically mature ova. A human female is born with a complement of about 2 million oocytes, and no more oocytes are produced postnatally.[17] During fetal life, all these cells begin the first meiotic division, but they become arrested at prophase. Very few (less than 450) of these oocytes complete meiotic maturation in adult life and undergo ovulation. Those that are ovulated develop within Graafian follicles that are stimulated to grow by follicle-stimulating hormone (FSH), secreted by the anterior pituitary. Usually only one of these follicles develops completely in each menstrual cycle. Approximately 14 days after the onset of menstrual flow, the anterior pituitary secretes a surge of luteinizing hormone (LH), which causes the resumption of meiosis in the oocyte within the large Graafian follicle. Approximately 24 hours later, the ovum is mature, having reached the metaphase II stage of the second meiotic division. At this time the Graafian follicle ruptures, and the ovum is ovulated and enters the upper end of the oviduct.

Three methods have been used over the years to obtain mature human ova. The earliest was based on the observations of Pincus and Enzmann[18] and Pincus and Saunders[19] that rabbit and human oocytes would spontaneously mature in vitro independent of hormonal stimulation. Edwards et al.[20] were unable to fertilize human ova produced in this way, and the approach was abandoned. The second approach, introduced by Steptoe and Edwards,[16] is to stimulate the production of mature ova artificially by the injection of gonadotropins. This procedure is usually performed with the injection of human postmenopausal gonadotropin (Pergonal), which is rich in FSH, followed by human chorionic gonadotropin, which is rich in LH. Steptoe and Edwards recovered ova from the Graafian follicle about four hours before the expected time of ovulation, initially by laparotomy and later by laparoscopy. In January 1979, Steptoe and Edwards summarized their work in an oral presentation in London before the Royal College of Obstetricians.[21] Of 77 patients who were treated with the injection of gonadotropins for ovum recovery, only three became pregnant after in vitro fertilization and embryo transfer; none of these pregnancies lasted for any length of time. This procedure was also abandoned. The third procedure is to allow a Graafian follicle to mature naturally without hormonal intervention.[1] Steptoe and Edwards detected the natural LH surge by the H-Gonavis test on the patient's urine. The time of this surge was used to predict the expected time of ovulation; ovum recovery was then done by laparoscopy a few hours before ovulation was due. These results were also reported to the Royal College, and they are summarized in table 6–1. Of 68 patients accepted for treatment, 45 yielded ova, which were exposed to spermatozoa in vitro; 32 of these embryos cleaved; four became implanted and resulted in established pregnancies, but only two of the implanted embryos resulted in newborn babies. Thus, as a rough estimate, the probability of producing a live birth with present techniques once a patient has been selected for ovum recovery is about 0.029. Should an ovum be recovered, the probability of obtaining a live birth is approximately 0.044.

The fertility of sexually active women without contraception can be described in terms of "effective fecundability." This term is defined as the probability of conception during a month resulting in a live birth.[22] Several attempts to estimate effective fecundability have been made. After a critical review, Leridon[23] concluded that the effective fecundability of women 25 years old is in the range of 0.20 to 0.25. At this level of effective fecundability, about four months of sexual activity is necessary for half the women to become pregnant, and eight months for 70 per cent to become pregnant (figure 6–1). The probability of attaining pregnancy with the present methods of Edwards and Steptoe is much lower than the natural effective fecundability, and thus the chances of achieving pregnancy, even with repeated operations, are extremely small.

Table 6–1
Results of In Vitro Fertilization and Embryo Transfer in Patients with Unstimulated Donors

Category	*Number of Patients*
Admitted to study	79
Sent home	11
Undergoing laparoscopy	68
Rejected because of adhesions	3
No ovulatory egg was recovered through errors of assessment of LH surge[a]	9
Egg could not be recovered from the follicle aspirate	11
Failure to fertilize egg in vitro	10
Failure of fertilized egg to cleave	3
Failure of transplanted egg to become implanted	28
Pregnancies	4
Babies born	2

Source: P.C. Steptoe and R.G. Edwards, "Report on Human In Vitro Fertilization and Embryo Transfer" (Presented before the Royal College of Obstetrics and Gynecology, London, 26 January 1979).

[a]LH denotes luteinizing hormone.

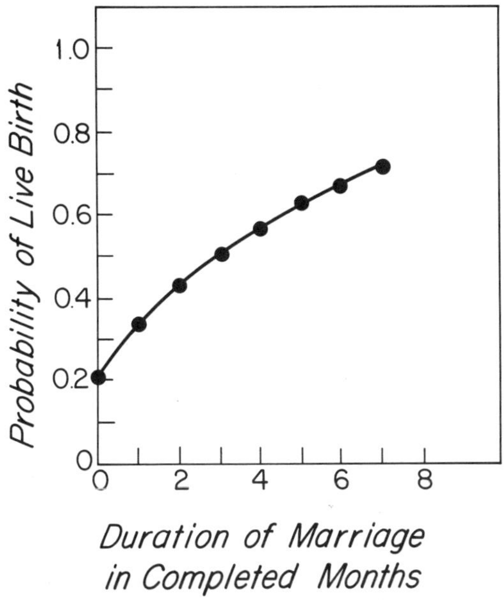

Source: The graph is calculated from the distribution-density function fitted to data from Tourouvre au Perche by J. Bongaarts, "A Method for the Estimation of Fecundability," *Demography* 12 (1975):645–660.

Figure 6–1. Probability of Live Birth after Different Completed Months of Marriage in a Population Not Practicing Birth Control

Can the Efficiency of the Procedure Be Increased?

To answer this question it is necessary to examine the incidence of natural embryonic and fetal wastage and its relation to effective fecundability.

In 1975, Roberts and Lowe[24] wrote an article in the *Lancet* with the intriguing title, "Where Have All the Conceptions Gone?" Their analysis showed that a high rate of embryonic loss occurs in women during the normal reproductive process. A similar phenomenon has been known for many years to occur in nonprimates. From studies in rabbits and pigs, Hammond[25] suggested in 1914 that embryonic loss is normal in mammalian reproduction. The widespread occurrence of this phenomenon was pointed out again by Robinson,[26] who was honored in 1921 with the invitation to give the Sir John Struthers Lecture of the Royal College of Surgeons of Edinburgh. The terms of this lectureship state that the recipient must not discuss problems in pathology. Robinson, who wanted to discuss embryonic death, clearly avoided the restriction in Struthers' will by arguing that although the death of a particular embryo involves pathologic processes, embryonic death is so widespread in mammals that it should be accepted as a normal phenomenon.

Hertig and Rock[27] first recognized early embryonic loss in women on the basis of the observation that potentially abortive ova can be recovered from women who are considered to be normally fertile. Only recently has it been recognized that this loss of embryos may be high, particularly in the early stages of pregnancy.[23,28,29] Estimates of the embryonic wastage in women are shown in figure 6–2.

The data can be used to estimate the probabilities of conception, recognizable pregnancy, and live birth in women who are reproducing naturally. After exposure to spermatozoa, the probability of fertilization of an ovum is 0.84 (table 6–2). By the time pregnancy is recognizable, half the embryos have been lost. During the remainder of pregnancy, another quarter of the embryos perish and are aborted. The entire process results in a

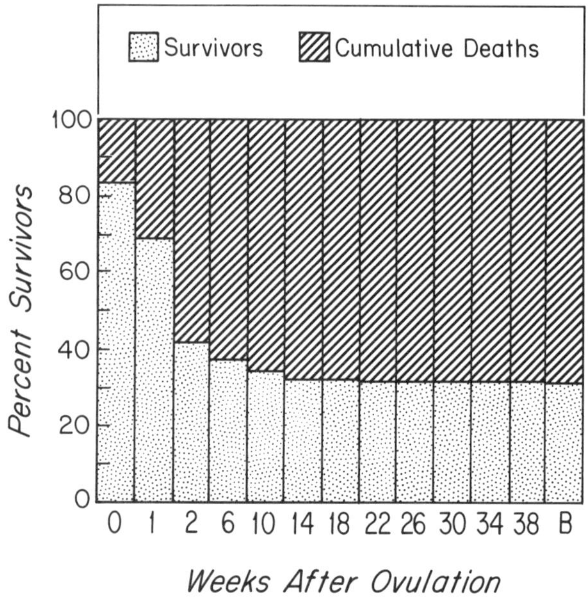

Source: The figures are based on the life table of H. Leridon, *Human Fertility* (Chicago: University of Chicago Press, 1977) and calculated from the data of A.T. Hertig and J. Rock, "A Series of Potentially Abortive Ova Recovered from Fertile Women Prior to the First Missed Menstrual Period," *American Journal of Obstetrics and Gynecology* 58 (1949):968–993; and F.E. French and J.M. Bierman, "Probabilities of Fetal Mortality," *Public Heath Report* 77 (1962):835–847.

Figure 6–2. Percentage of Surviving and Lost Human Embryos and Fetuses at Different Stages of Pregnancy

Table 6–2
Estimates of the Probability of Conception, Recognizable Pregnancy, and Live Birth by Natural Reproduction and by In Vitro Fertilization and Embryo Transfer

	Natural		*in Vitro*	
Result	*Estimate*	*Proportion Lost*	*Estimate*	*Proportion Lost*
Conception	0.84[a]	—	0.78[b]	—
Recognizable pregnancy	0.42	0.50[c]	0.09	0.88[c]
Live birth	0.31	0.26[d]	0.04	0.50[d]

Sources: Data for natural reproduction from H. Leridon, *Human Fertility* (Chicago: University of Chicago Press, 1977). Data for in vitro fertilization and embryo transfer from P.C. Steptoe and R.G. Edwards, "Report on Human In Vitro Fertilization and Embryo Transfer" (Presented before the Royal College of Obstetrics and Gynecology, London, 26 January 1979).

[a]Leridon's data showed that 16 of 100 ova are not normally fertilized.

[b]The probability of obtaining a fertilized ovum (conception) is given by the ratio of the number of ova fertilized (35) to the number of patients from whom ova were recovered (45).

[c]Lost by the time pregnancy would have been detected.

[d]Lost during remainder of pregnancy.

probability of a live birth of only 0.31. This value is slightly greater than the estimates derived from studies of the time to first birth after marriage in a population not practicing contraception.[22] The difference may be due to failure of coitus or sperm transport. Similar probabilities can also be estimated from the data of Steptoe and Edwards (table 6–1) on the results of in vitro fertilization and embryo transfer in women. The probability of conception after in vitro fertilization is similar to that observed in vivo, so fertilization does not seem to limit the effectiveness of Steptoe and Edwards' technique. However, with this method a larger proportion of the embryos are lost by the time pregnancy is recognizable. Since fertilized human ova cleave readily in culture,[30] the much greater embryonic loss after in vitro fertilization and embryo transfer is probably associated with implantation after introduction into the patient's uterus. The results in table 6–2 also suggest a large loss of embryos in later pregnancy after in vitro fertilization and embryo transfer, but since only four pregnancies have been achieved, the number of cases is too small to yield reliable conclusions. Thus, if we are to improve the effectiveness of in vitro fertilization and embryo transfer, research will be needed on the period of implantation in women. The other area requiring further research concerns the collection of mature ova from the Graafian follicle.

Even if the collection of ova and the transfer technique could be completely perfected, the resulting performance would be no more efficient than the normal process (figure 6–1). Thus, if a single embryo is produced and transferred, the procedure may have to be repeated about four times for pregnancy to occur in 50 per cent of the patients, and eight times for success in 70 per cent. On the basis of experience with laboratory and domestic animals (where transfer is into surrogate mothers) the procedure is 25 to 50 per cent efficient. In women with an effective fecundability of 0.31, an efficiency of 35 per cent implies that the probability of conception is only 0.11.

Repetition of the procedure until pregnancy is attained is clinically unacceptable if general anesthesia has to be used monthly for ovum recovery. The problem might be eased if several oocytes could be recovered at a single operation and stored in a frozen condition for use in successive cycles. Discarding unfertilized oocytes should the patient become pregnant in the first or second cycle would not raise ethical objections, as does discarding preimplantation embryos. Mouse oocytes have already been stored in the frozen state, thawed, fertilized in vitro, and cultured; after transfer to uterine foster mothers, normal fetuses and newborns developed.[31]

Another method of increasing efficiency is to produce and transfer more than one embryo. This method, however, introduces the possibility of multiple births. The probabilities of multiple births after transfer of two and three blastocysts, if one assumes that the probability of a single birth is 0.11, can be calculated with the binomial distribution. The transfer of two embryos raises the probability of success to 0.21, with a chance of 1 per cent of patients having twins. The transfer of three embryos increases the probability of success to 0.29, with a 3 per cent chance of twins and a 1 in 1000 chance of triplets.

What Is the Probability of Producing a Congenitally Abnormal Baby?

To answer this question it is necessary to review the causes of embryonic and fetal wastage. The causes fall into two classes. One class consists of intrinsic abnormalities of the embryo. These disorders may be unmasked recessive genes such as homozygous alpha-thalassemia or chromosomal aberrations such as triploidy, in which the embryo contains three complements of the haploid set of chromosomes. The other class consists of lethal environmental factors in the female genital tract. These factors may originate from overt disease of the genital tract, from the presence of such endogenous teratogenic agents as the mumps and rubella viruses,[32] or from the normal aging of the female genital tract, resulting in impairment of implantation and placental function.[33] Cytogenetic studies of spontaneous human abortuses show that chromosomal abnormalities are a major cause of embryonic and fetal loss. Recently, Schlesselman [34] reviewed the available literature and concluded that 40 to 50 per cent of human embryos that are successfully implanted may have a chromosome abnormality that leads to their loss before birth. The incidence of these cytogenetic aberrations tends to be higher in older women.[23,35] The frequency of the noncytogenetic causes of embryonic and fetal wastage is unknown.

There are two main classes of chromosomal aberration found in human embryos. One class consists of the monosomies and trisomies. The two meiotic reduction divisions that occur in oogenesis and spermatogenesis normally result in one member of each of the 23 pairs of chromosomes passing at random into the ovum or spermatozoon. Should this separation fail for a particular chromosome pair, the gamete will possess either the two chromosomes or none. If these gametes fuse with normal gametes at fertilization, the embryo will either have only one of a particular chromosome (instead of two) or it will have three chromosomes. The former type of aberration, a monosomy, occurs in Turner's syndrome, for example, in which either an X or Y chromosome is missing. The latter type, a trisomy, occurs in Klinefelter's syndrome, in which the embryo has an XXY constitution. Another example of a trisomy is Down's syndrome, in which Chromosome 21 is represented three times. The second main class of chromosomal aberrations is polyploidy, in which more than two multiples of the entire haploid set of chromosomes occur. The commonest, triploidy, occurs in about 2 per cent of clinically recognized conceptions.[36] This condition can arise either by the fertilization of a normal haploid ovum by more than one sperm, by the fertilization of a haploid ovum with a diploid sperm, or by the fertilization of a diploid ovum with a haploid sperm. Diploid gametes appear through errors of meiosis, in which retention of the full 23 pairs of chromosomes occurs.

The presence of abnormal complements of chromosomes does not prevent early embryonic development. However, very few of these embryos are capable of development to term, and they are eliminated as spontaneous abortions. Some conditions are rapidly lethal: embryos with trisomy of the E class of chromosomes, most monosomies, or tetraploidy are eliminated by the sixth week. Other conditions, such as trisomies C, D, and G, triploids, and the majority of cases of monosomy X, cause death by the 10th week. In all these syndromes, development ceases because of the lack of proper genetic information that controls development, not because of influences from the mother.

The quality of the embryos produced in vitro will depend, as in those produced naturally in vivo, on the quality of the gametes that meet at fertilization. Since the normality of the spermatozoa and ova used for in vitro fertilization cannot be controlled, it is inevitable that a considerable proportion of chromosomally abnormal embryos will arise and that they will be transferred back into the infertile patients. It is to be hoped that most of them will be eliminated in early pregnancy, as occurs under normal conditions. Nevertheless, a few of these embryos may be born and may even have some form of congenital anomaly, as also occurs naturally. This fact of nature has highly important legal implications should damages be sought in the case of an abnormal baby born after in vitro fertilization and embryo transfer.

A more specific question is whether the technique of in vitro fertilization and embryo transfer will increase the probability of birth defects to above the normal level. An increase in the incidence of abnormalities could occur in four ways: the induction of chromosomal aberrations, an increase in the rate of fertilization by abnormal spermatozoa, the induction of point mutations, and the actions of physical and chemical teratogens. Boué and Boué[37] reported an increase in the level of trisomy among abortuses from women who had been induced to ovulate. Since then, several reports have confirmed that an increase in the incidence of chromosomal aberrations occurs as the result of superovulation in rabbits[38,39] and in mice.[40,41] The manner in which ova are exposed to spermatozoa in vitro may also result in an increase in triploidy. If high concentrations of spermatozoa are used, the natural block to polyspermy may be overwhelmed, resulting in polyspermic fertilization. In mice, an increase in triploidy of 12 per cent has been observed.[42,43] Edwards and Steptoe[30] stated that they examined the chromosomes of 15 human blastocysts produced in vitro and found no instance of triploidy. This observation is consistent with a normal incidence of 2 per cent triploidy in all known conceptions.[36] However, the sample is too small to yield a definitive conclusion. Edwards and Steptoe believed that their chromosome counts were too inaccurate to show that the blastocysts were normal diploids; the lack of reliable data on this point is unfortunate.

There is evidence that some types of abnormal sperm are eliminated during the passage through the female genital tract. For example, it has been shown in women that few morphologically abnormal sperm reach the site of fertilization.[44] Similar observations have been made in mice, and it has been suggested that the uterotubal junction is a barrier to abnormally shaped spermatozoa.[45] Studies of a different type (Bedford MJ, Personal communication) have shown that after spermatozoa have undergone final maturation in the oviduct (a process called capacitation) they have a fertilizing capacity much greater than that of those recovered from the uterus.[46] Thus, fertilization of ova in vitro with spermatozoa that have not passed through the female genital tract may be associated with an increased risk of fertilization with abnormal spermatozoa. The extent of this risk is an open question.[47]

There is no experimental evidence that bears on the possible contributions of induced point mutations and of teratogenic effects that could be attributed directly to the techniques of in vitro fertilization and transfer of embryos. None of the methods involves the use of known mutagenic or teratogenic agents. Moreover, the preimplantation embryo is highly insensitive to teratogens, compared with the sensitivity that develops at the time of organogenesis.[48] For example, several studies have shown that high doses of a highly teratogenic agent may kill the preimplantation embryos before blastocyst formation or may reduce the rate of cleavage. As a result, there are insufficient cells to form the inner cell mass, and a simple vesicle forms that is incapable of giving rise to an embryo.[49] Other teratogenic agents may also cause embryonic death instead of causing anomalies affecting specific organs later in development.

If freezing of human preimplantation stages is used in the future to hold embryos for later transfer, the effects of low temperature may cause concern. This question has already been raised over the low-temperature preservation of early embryos of genetically valuable mouse strains to avoid the expense of large breeding colonies. There is no evidence that freezing itself is mutagenic.[50] However, it has been suggested that the cumulative effect of background radiation may increase the mutation rate, since no DNA repair will occur at the temperature of liquid nitrogen. Recent studies by Whittingham et al.[51] have shown that 84 times the background radiation of frozen mouse embryos for 27 to 29 months produced no detectable abnormal offspring.

A considerable amount of experience has accumulated with laboratory and domestic animals on in vitro fertilization and embryo culture, employed separately or in sequence, followed by transfer to surrogate mothers.[52–54] The combined evidence gives no indication that these techniques cause an increase in the number of abnormal offspring. The issue of whether sufficient work has been done in primates has led to considerable debate.[55,56]

To summarize, there is evidence that the technique of in vitro fertilization and embryo transfer may increase the incidence of cytogenetic defects in early embryos. If the ova are obtained without the use of superovulation, these abnormal embryos will be mainly polyploidies. When transferred back into the patient's uterus, almost all these embryos can be expected to die, as they do in a natural pregnancy. There is little or no compelling evidence, either experimental or theoretical, suggesting that other types of abnormality will be increased.

Conclusion

The success of Steptoe and Edwards has attracted worldwide publicity, with the result that the hopes of many infertile couples may have been raised excessively. The technique is still imperfect, and even if its efficiency is raised considerably through future research, the procedure may have to be repeated several times to achieve success in even 50 per cent of patients. Thus, the procedure will only be used as a last resort after all other treatments have failed. Nevertheless, the procedure may be useful in cases in which couples desire to produce their own child rather than to adopt, or in countries where few babies are available for adoption. The evidence suggests that the danger of increased congenital defects is not high. The risk seems considerably lower than that accepted by couples with a recessive defect who decide to have a child even though they know that an abnormal baby may be born.

References

1. P.C. Steptoe and R.G. Edwards, ''Birth after the Reimplantation of a Human Enbryo,'' *Lancet* 2 (1978):366.
2. R. Edwards and P. Steptoe, *A Matter of Life* (London: Hutchinson, 1980).
3. J.D. Biggers, ''Fertilization and Blastocyst Formation,'' in N.J. Alexander ed., *Animal Models for Research on Contraception and Fertility* (Hagerstown, Md.: Harper & Row, 1979), pp. 223–237.
4. J.P. Greenhill, ''Evaluation of salpingostomy and tubal implantation for the treatment of sterility,'' *American Journal of Obstetrics and Gynecology* 33 (1937):39–51.
5. R.T. Morris, ''The ovarian graft,'' *New York Medical Journal* 62 (1895):436–437.
6. C.E. Adams, ''Consequences of accelerated ovum transport, including a re-evaluation of Estes' operation,'' *Journal of Reproduction and Fertility* 55 (1979):239–246.
7. Anonymous, ''Conception in a Watch Glass,'' *New England Journal of Medicine* 217 (1937):678.
8. C.R. Austin and M.W.H. Bishop, ''Fertilization in mammals,'' *Biological Review* 32 (1957):296–349.
9. W. Heape, ''Preliminary note on the transplantation and growth of mammalian ova within a uterine foster mother,'' *Proceedings of the Royal Society of London* 48 (1890):457–458.
10. J. Hammond, Jr., ''Recovery and culture of tubal mouse ova,'' *Nature* 163 (1949):28–29.
11. W.K. Whitten, ''Culture of tubal mouse ova,'' *Nature* 177 (1956):96.
12. *Idem*. ''Culture of tubal ova,'' *Nature* 179 (1957):1081–1082.
13. A. McLaren and J.D. Biggers, ''Successful development and birth of

mice cultivated in vitro as early embryos,'' *Nature* 182 (1958): 877–878.

14. M.C. Chang, ''Fertilization of rabbit ova in vitro,'' *Nature* 184 (1959):466–467.
15. R.G. Edwards, ''Maturation in vitro of human ovarian oöcytes,'' *Lancet* 2 (1965):926–929.
16. P.C. Steptoe and R.G. Edwards, ''Laparoscopic recovery of preovulatory human oocytes after priming of ovaries with gonadotrophins,'' *Lancet* 1 (1970):683–689.
17. T.G. Baker, ''Oogenesis and ovulation,'' in C.R. Austin, R.V. Short eds., *Reproduction in Mammals,* 1 (Cambridge: Cambridge University Press, 1972), pp. 14–45.
18. G. Pincus and E.V. Enzmann, ''The comparative behavior of mammalian eggs in vivo and in vitro. I. The activation of ovarian eggs,'' *Journal of Experimental Medicine* 62 (1935):665–675.
19. G. Pincus and B. Saunders, ''The comparative behavior of mammalian eggs in vivo and in vitro,'' *Anatomical Record* 75 (1939):537–545.
20. R.G. Edwards; R.P. Donahue; T.A. Baramki; and H.W. Jones, Jr., ''Preliminary attempts to fertilize human oocytes matured in vitro,'' *American Journal of Obstetrics and Gynecology* 96 (1966):192–200.
21. P.C. Steptoe and R.G. Edwards, ''Report on human in vitro fertilization and embryo transfer'' (Presented before the Royal College of Obstetrics and Gynecology, London, 26 January 1979).
22. J. Bongaarts, ''A method for the estimation of fecundability,'' *Demography* 12 (1975):645–660.
23. H. Leridon, *Human Fertility* (Chicago: University of Chicago Press, 1977).
24. C.J. Roberts and C.R. Lowe, ''Where have all the conceptions gone?'' *Lancet* 1 (1975):498–499.
25. J. Hammond Sr., ''On some factors controlling fertility in domestic animals,'' *Journal of Agricultural Science* (Camb.) 6 (1914):263–277.
26. A. Robinson, ''Prenatal death,'' *Edinburgh Medical Journal* 26 (1921):137–151, 209–231.
27. A.T. Hertig and J. Rock, ''A series of potentially abortive ova recovered from fertile women prior to the first missed menstrual period,'' *American Journal of Obstetrics and Gynecology* 58 (1949):968–993.
28. H. Nishimura, ''Fate of human fertilized eggs during prenatal life: present status of knowledge,'' *Okajimas Folia Anatomica Japanica* 46 (1970):297–305.
29. F.E. French and J.M. Bierman, ''Probabilities of fetal mortality,'' *Public Health Report* 77 (1962):835–847.
30. R.G. Edwards and P.C. Steptoe, ''The relevance of the frozen storage of human embryos,'' *Ciba Foundation Symposium* 52 (1977):235–243.

31. D.G. Whittingham, "Fertilization in vitro and development to term of unfertilized mouse oocytes previously stored at −196°C," *Journal of Reproduction and Fertility* 49 (1977):89–94.
32. M. Siegel; H.T. Fuerst; and N.S. Peress, "Comparative fetal mortality in maternal virus diseases: A prospective study on rubella measles, mumps, chicken pox and hepatitis," *New England Journal of Medicine* 274 (1966):768–771.
33. J.D. Biggers, "Problems concerning the uterine causes of embryonic death, with special reference to the effects of ageing of the uterus," *Journal of Reproduction and Fertility*(Suppl.) 8 (1969):27–43.
34. J.J. Schlesselman, "How does one assess the risk of abnormalities from human in vitro fertilization?" *American Journal of Obstetrics and Gynecology* 135 (1979):48.
35. J.G. Boué and A. Boué, "Chromosomal anomalies in early spontaneous abortion: their consequences on early embryogenesis and in vitro growth of embryonic cells," *Current Topics in Pathology* 62 (1976):193–208.
36. P.A. Jacobs, "Population surveillance: a cytogenetic approach," in N. E. Norton and C.S. Chung, eds., *Genetic Epidemiology* (New York: Academic Press, 1978), pp. 563–581.
37. J.G. Boué and A. Boué, "Increased frequency of chromosomal anomalies in abortions after induced ovulation," *Lancet* (1973):679–680.
38. S. Fujimoto; N. Pahlavan; and W.R. Dukelow, "Chromosome abnormalities in rabbit preimplantation blastocysts induced by superovulation," *Journal of Reproduction and Fertility* 40 (1974):177–181.
39. S. Fujimoto; T.J. Passantino; and I. Koenczoel, "A preliminary note on chromosome abnormalities in intratubal rabbit embryos," *Proceedings Japanese Academy* 51 (1975):51–55.
40. N. Takagi and M. Sasaki, "Digynic triploidy after superovulation in mice," *Nature* 264 (1976):278–281.
41. I. Maudlin and L.R. Fraser, "The effect of PMSG dose on the incidence of chromosomal anomalies in mouse embryos fertilized in vitro," *Journal of Reproduction and Fertility* 50 (1977):275–280.
42. L.R. Fraser; H.M. Zanellotti; G.R. Paton; and L.M. Drury, "Increased incidence of triploidy in embryos derived from mouse eggs fertilised in vitro," *Nature* 260 (1976):39–40.
43. L.R. Fraser and I. Maudlin, "Relationship between sperm concentration and the incidence of polyspermy in mouse embryos fertilized in vitro," *Journal of Reproduction and Fertility* 52 (1978):103–106.
44. M. Ahlgren, "Sperm transport to and survival in the human fallopian tube," *Gynecological Investigations* 6 (1975):206–214.

45. H. Krzanowska, "The passage of abnormal spermatozoa through the uterotubal junction of the mouse," *Journal of Reproduction and Fertility* 38 (1974):81–90.
46. J. Cohen, "Gametic diversity within an ejaculate," in B. Afzelius, ed., *The functional anatomy of the spermatozoon* (New York: Pergamon Press, 1975), pp. 329–339.
47. J.W. Overstreet and D.F. Katz, "Sperm transport and selection in the female genital tract," in M.H. Johnson, ed., *Development in mammals*, vol. 2 (Amsterdam: North-Holland, 1977), pp. 31–65.
48. F. Beck and J.B. Lloyd, "Embryological principles of teratogenesis," in J.M. Robson, F.M. Sullivan, R.L. Smith, eds., *Embryopathic activity of drugs* (Boston: Little, Brown, 1965), pp. 1–20.
49. H. Alexandre, "Correlation between cytological and morphogenetic effects of acute X-irradiation of preimplantation mouse embryos cultured in vitro," *Journal of Reproduction and Fertility* 53 (1978): 399–402.
50. P. Mazur, "Freezing and low temperature storage of living cells," in O. Mühlbock, ed., *Basic aspects of freeze preservation of mouse strains* (Stuttgart: Gustav-Fischer Verlag, 1975), pp. 57–65.
51. D.G. Whittingham; M.F. Lyon; and P.H. Glenister, "Long-term storage of mouse embryos at −196°C: The effect of background radiation," *Genetic Research* 29 (1977):171–181.
52. J.D. Biggers, "In vitro fertilization, embryo culture and embryo transfer in the human" (Prepared for the Ethics Advisory Board of the U.S. Department of Health, Education, and Welfare, 15 September 1978.)
53. C. Rice, Supplement to note 52.
54. R.H. Foote, "In vitro fertilization in perspective, relative to the science and art of domestic animal reproduction" (Prepared for the Ethics Advisory Board, U.S. Department of Health, Education, and Welfare, 20 October 1978.)
55. Ethics Advisory Board, Department of Health, Education, and Welfare, "Report and conclusions: HEW support of research involving human in vitro fertilization," 4 May 1979.
56. R.J. Blandau, "In vitro fertilization and embryo transfer," *Fertility and Sterility* 33 (1980):3–11.

Addendum

John D. Biggers

Since chapter 6 was written, further progress has been reported. Some of the work in Australia has been described by Lopata et al.,[1] Trounson et al.,[2] and Wood et al.[3] At present, sixteen or seventeen babies have been born, mainly in England and Australia, following in vitro fertilization and embryo transfer although the details of many of the cases have not been described. In one delivery, twins were born. Several other women are currently pregnant following the procedure, including two in the United States.

With one exception, all births have resulted in normal babies. The exception occurred in the case that resulted in twins. One sibling suffered from a correctible congenital heart defect. There is no reason to conclude that this congenital defect was caused by the techniques of in vitro fertilization and embryo transfer since such cases arise under natural conditions.

In a recent issue of the *Obstetrical and Gynecological Survey,* an editorial note quotes some recent data on in vitro fertilization and embryo transfer obtained by Wood and his colleagues at the Queen Victoria Medical Center in Melbourne, Australia:[4]

Patients undergoing laparoscopy:	152
Number of ova recovered:	130
Number of ova fertilized:	103
Number of embryos transferred:	91
Number of pregnancies:	15

These results can be compared statistically with the expected results predicted by Leridon's life table (figure 6–2) for reproduction occurring naturally. There are no significant differences between the fertilization rate and the rate of preimplantation development under in vitro and in vivo conditions. In contrast, there is a highly significant difference ($P < .001$) between the number of pregnancies produced. Embryo transfer resulted in only a 16.5 percent success rate compared with 53.6 percent under natural conditions.

Recent discussions of the ethical and legal aspects of in vitro fertilization and embryo transfer for the treatment of human infertility have been written by Grobstein and Evans and Dixler respectively.[5]

Notes

1. A. Lopata; J.B. Brown; J.R. Leeton; J.McTalbot; and C. Wood, ''In Vitro Fertilization of Preovulatory Oocytes and Embryo Transfer in Infertile Patients Treated with Clomithene and Human Chorionic Gonadotropin,'' *Fertility and Sterility* 30 (1978):27–35.

2. A.O. Trounson; J.R. Leeton; C. Wood; J. Webb; and J. Wood, ''Pregnancies in Humans by Fertilization *in vitro* and Embryo Transfer in the Controlled Ovulatory Cycle,'' *Science* 212 (1981):681–682.

3. C. Wood; A.O. Trounson; J.R. Leeton; J.McTalbot; B. Buttery; J. Webb; J. Wood; and D. Jessup, ''A Clinical Assessment of Nine Pregnancies Obtained by *in vitro* Fertilization and Embryo Transfer,'' *Fertility and Sterility* 35 (1981):502–508.

4. Editorial note, ''Wood et al.: A Clinical Assessment of Nine Pregnancies Obtained by *in vitro* Fertilization and Embryo Transfer,''*Obstetrical and Gynecological Survey* 36 (1981):455–456.

5. C. Grobstein, *From Chance to Purpose* (Reading, Mass.: Addison-Wesley, 1981); and M.I. Evans and A.O. Dixler, ''Human *in vitro* Fertilization. Some Legal Issues,'' *Journal of the American Medical Association* 245 (1981):2324–2327.

7 In Vitro Fertilization and Embryo Transfer: Public Policy and Ethics

Warren T. Reich

What should U.S. federal policy say about in vitro fertilization and embryo transfer, both in the research setting and in clinical practice? What ethical considerations should mold the public policy that governs these activities? To appreciate the context in which these questions are asked, as well as the reflections I offer, it is important to be aware of three milestone events that had a great impact on the formation of public policy for in vitro fertilization.

Events Shaping In Vitro Fertilization Policy Debates

The arrival of "the world's first test-tube baby," to use the language of a U.S. newspaper's headlines, had an enormous impact on both the public itself and public policymakers. I am referring, of course, to Louise Brown, who was born on 25 July 1978 to Lesley Brown, 31, and Gilbert Brown, 38, in Oldham, England. An apparently normal, healthy girl weighing 2,700 g (5 lbs, 12 ozs), the infant was the world's first baby known to be successfully delivered after being conceived outside the human body. Dr. Patrick Steptoe, an obstetrician-gynecologist at the Oldham General Hospital, and Dr. Robert G. Edwards, a Cambridge University physiologist, circumvented Mrs. Brown's problem of infertility, which was due to blockage of the Fallopian tubes, by fertilization of Mrs. Brown's egg with Mr. Brown's semen in vitro, followed by implantation of the new, eight-cell embryo into the mother's uterus two-and-one-half days after fertilization. Approximately eight-and-one-half months later, baby Brown was born.[1]

Several aspects of the Brown case were striking. First, from a medical perspective, baby Brown was and still is apparently healthy; she is not a visible mockery of human biological manipulation. (This does not, of course, settle the question of whether in vitro fertilization poses an accept-

---

This chapter was first delivered as a lecture at the Science and Morality Conference at Nazareth College, Rochester, N.Y., 5 April 1979. It was revised in August 1981 for publication in this volume. Research for this project was supported in part by the National Endowment for the Humanities, Grant No. ED-32239-78-909 to the Program in Health and Humanities, Georgetown University. I wish to thank Laurence McCullough for making helpful suggestions to improve the text.

able risk of defect in the child to be generated in this way—a question that is discussed later.) Second, the Brown case helped to create a state of the art that now increasingly makes possible the application of in vitro fertilization for circumventing infertility caused by tubal pathological conditions.[2] Third, it was immediately apparent that, aside from the limited ethics publications cited in this chapter, widespread public debate on this topic unfortunately had not preceded the advent of in vitro fertilization. Thus, policymakers began their policymaking efforts in a sensational atmosphere characterized by headlines over two newspaper columns in the *Washington Post:* "Irreverent Test Tubes"[3] and "Man-Made Evolution."[4]

A second event significant for in vitro fertilization public policy was a legal suit brought by John Del Zio, a Florida dentist, and his wife, in which they charged that in September 1973, Dr. Raymond Vande Wiele, director of obstetrical and gynecological services at Columbia-Presbyterian Medical Center in New York, deliberately destroyed a culture containing ova of Mrs. Del Zio without consent of the would-be parents, one day after they had been fertilized with her husband's semen by Dr. Landrum Shettles and three days before the human blastocyst was to be implanted in Mrs. Del Zio's womb. The suit, which named Vande Wiele, Presbyterian Hospital, and the trustees of Columbia University as defendants, asked $1.5 million in damages. By coincidence, the trial was concluded in a federal court five years later, just one month after the world's first in vitro fertilization–conceived baby (Louise Brown) was born in 1978.[5] Dr. Vande Wiele defended his action on grounds that the procedure had been performed unethically, that it was undertaken by unqualified physicians and personnel, and that Mrs. Del Zio, following the embryo transplant, would possibly have contracted peritonitis (which might have proved fatal to her). The jury awarded $50,000 to the would-be parents for the severe emotional distress they had suffered. However, the defendants were found innocent of wrongfully converting property belonging to Mrs. Del Zio. Conversion is the wrongful taking of the personal property of another person.[6] Under this theory, the Del Zios had argued that the contents of the test tube (the developing embryo) were their property and not a person. Because of the publicity surrounding this case, it undoubtedly influenced public reflection on policies that should govern in vitro fertilization.[7]

A third and more pervasive element creating the milieu of the policy debates on federal in vitro fertilization policy was the *Roe* v. *Wade* abortion decision of the U.S. Supreme Court in 1973.[8] The majority opinion of this well-known decision declared that the fetus is not a person within the language and meaning of the Fourteenth Amendment and argued that in most areas the law has been "reluctant to endorse any theory that life, as we recognize it, begins before live birth or to accord legal rights to the unborn except in narrowly defined situations and except when the rights are contingent upon live birth."[9] The Court found more convincing than the rights or claims of the fetus the pregnant woman's right to privacy, but qualified her right to privacy by declaring that the state could have a compelling state interest in potential life beginning with the point of viability; and, consequently, that the individual state may proscribe the abortion of viable fetuses, "except when it is necessary to preserve the life or health of the mother."[10]

There is one significant difference between the questions raised by the *Roe* v. *Wade* decision and the in vitro fertilization federal policy decision—namely, while the former deals with the postimplantation fetus, the latter deals primarily with the preimplantation embryo. Yet they share one large ethical question: What are the moral and legal claims of the early, prenatal stages of human life? It was only a matter of months following the *Roe* v. *Wade* decision that opposition to the Supreme Court's permissive abortion policies accounted for strong congressional agitation against reported instances of live-fetus research—an opposition that was to have an effect on in vitro fertilization research policy.[11]

In the highly charged atmosphere typified by the Brown in vitro breakthrough, the Del Zio civil suit, and the U.S. Supreme Court's abortion decision, public officials had to face the question of the formulation of federal policy on in vitro fertilization and embryo transfer in light of a variety of conflicting ethical and legal considerations.

Federal Policy on In Vitro Fertilization Research and Embryo Transfer

Attempts to establish federal laws and regulations regarding research with human in vitro fertilization have involved both the U.S. Congress and the Department of Health, Education, and Welfare (DHEW) and have been closely linked with the policymaking process for fetal research.

In May 1973, the U.S. Congress, reacting to reports regarding experiments involving live human fetuses, and with the U.S. Supreme Court decision on abortion *(Roe* v. *Wade)* still fresh on their minds, considered proposed legislation that would have banned the use of federal funds for live-fetus research.[12] On 16 November 1973, the DHEW issued its first of three documents regarding guidelines for research involving in vitro fertilization and/or embryo transfer—a "draft working document of proposed policy."[13] Following some fascinating congressional floor debates on fetal research and in vitro fertilization,[14] the U.S. Congress, in July 1974, voted a temporary ban on certain types of fetal research and created a national commission to study the issue of research with fetuses and other human subjects and to report its findings to the secretary of the DHEW.[15] Before that commission (the National Commission for the Protection of Human Subjects of Biomedical and Behavioral Research) could hold its first meeting, the DHEW, in August 1974, published proposed rule making on research with several specific groups of human subjects; in this document the DHEW recommended that the board consider several scientific, ethical, and legal issues pertaining to in vitro fertilization and/or embryo transfer.[16]

It is interesting that the commission, which began its work in December 1974, was not given the charge of proposing regulations on in vitro fertilization, apparently because research with other human subjects, particularly fetuses, was regarded as a matter of greater public controversy. In 1975, the commission issued its report and recommendations for regulation of research on the human fetus,[17] recommending the permissibility of support of fetal research by the DHEW within detailed limits, against the background of its general conviction that "the fetus, regardless of life prospects, should be treated respectfully and with dignity."[18]

When the secretary of the DHEW subsequently published regulations regarding research involving fetuses and pregnant women in the *Federal Register* of 8 August 1975—regulations that took into account the recommendations of the national commission—he included in the regulations some stipulations regarding in vitro fertilization.[19] These provided that no DHEW funds could be used for research involving human in vitro fertilization until the application or proposal for the research project had been reviewed by an ethical advisory board and until that board had given advice as to the project's acceptability from an ethical standpoint. Since the Ethics Advisory Board was not appointed for another three years (in May 1978) and did not issue its report on in vitro fertilization until 1979, the effect of this review requirement was that there existed a four-year federal moratorium on research involving in vitro fertilization and embryo transfer. Although this moratorium covered only DHEW-sponsored research, many scientists were convinced that its effect was to create a virtually total ban on in vitro fertilization research.

The first case presented to the Ethics Advisory Board by DHEW Secretary Joseph Califano was a proposal that had been submitted in 1977 by Dr. Pierre Soupart of Vanderbilt University, who wanted to conduct laboratory studies on human embryos following in vitro fertilization. The Ethics Advisory Board conducted a thorough study of in vitro fertilization in both the laboratory research setting (where there is no intention of implanting the embryo in the uterus after the research project) and in the clinical setting (where embryo transfer would occur for purposes of achieving a pregnancy). In its final recommendations, submitted to Secretary Califano in May 1979, the board concluded that it is acceptable from an ethical standpoint to undertake research involving in vitro fertilization and embryo transfer under certain conditions, but it made no judgment on whether the DHEW should fund this research.[20]

Publication of the report in the *Federal Register* elicited close to 13,000 comments signed by 55,000 persons; 95 percent of this public comment was opposed to in vitro fertilization research. As of August 1981, no further policy decision has been made by the secretary of DHEW, called since May 1980 the Department of Health and Human Services (DHHS). (Dr. Soupart died before a decision could be made about the funding of his project by DHHS.)

Hence, the regulations remain the same: No research in in vitro fertilization and embryo transfer may be done under sponsorship of the DHHS until its ethical acceptability has been reviewed by the Ethics Advisory Board (which no longer exists) or by another appropriate committee of the public-health service. Of course, this research can be conducted with non-DHHS funds, if it is done at an institution that has a status of general assurance of compliance with appropriate codes of ethics and with the regulations of the National Institutes of Health (NIH), and if the research project has been approved by the institution's institutional review board or an equivalent body. Such is the case at the In Vitro Fertilization Clinic of the Eastern Virginia Medical School in Norfolk, Virginia, where in vitro fertilization and clinical embryo transfer are being carried out.[21]

Public-Policy Ethics: Some Criteria

The foregoing recent history of federal policies regarding in vitro fertilization and/or embryo transfer is of special interest because these policymaking processes and documents reveal information about the relationship of ethics to public policy in the area of in vitro fertilization and embryo transfer. Specifically, they help to clarify the objectives and methods employed in this area of health policy.

For example, it becomes apparent that there are a variety of ways in which public policy might act in reference to in vitro fertilization and embryo transfer: (1) *through direct restriction,* which could take any of three forms: in vitro fertilization could be totally banned; it could be limited to certain types, for example, to in vitro fertilization only when there is the intention to implant the zygote in the uterus; or it could be restricted to specific purposes, for example, in vitro fertilization research designed to establish the safety and efficacy of embryo transfer could be approved, as the Ethics Advisory Board did,[22] while research designed to test the effect of noxious stimuli on the human embryo could be forbidden; (2) *through refusal of government funding or sponsorship* of some or all in vitro fertilization research procedures, which would have the effect of discouraging the research and clinical practices in this area but not the effect of restricting nongovernmentally supported activities; or (3) *through procedural requirements,* for example, by stipulating that in vitro fertilization and/or embryo transfer research projects may not be undertaken or supported by a government agency until the proposal has been reviewed by an ethics advisory committee. Because these kinds of policy actions vary enormously, one cannot hastily conclude that regulations, laws, or other forms of public policy necessarily entail major infringements on human freedom. Yet the development of all the relevant types of public policy depends partly on presuppositions about the relationship of moral principles to public policy.

It is quite impossible properly to understand the debate about in vitro fertilization and embryo transfer as a public, social issue unless we give prior consideration to the question: What are the criteria for selecting the moral values and principles that should give shape to public policy for in vitro fertilization and embryo transfer?

At first glance, the question would not appear to be too difficult, considering the affinity of ethics to public policy.[23] Ethics can be defined as "the reflective study of what is good or bad in that part of human conduct for which man has some personal responsibility,"[24] and public policy consists "of decision rules adopted by authorities as a guide to behavior."[25]

Both ethics and public policy are concerned with the same question: What policy is good for the society? Yet they differ widely in both method and goal, for while ethics is concerned with the explication, rational analysis, and justification of moral principles regarding what is right and good and their general application to certain kinds of behaviors, the policymaking process "involves not just the logical movement of ideas, but the pressure of emotions, the weight of [political] interests, the bargaining and credit taking of authorities and constituencies."[26]

Perhaps the strongest conflict between ethics and public policy arises from the moral pluralism of modern society. In the pluralistic society, many ideologies and moral principles vie for acceptance, and policymakers are likely to be regarded as mediators of irreconcilable issues in their efforts to establish peaceable social arrangements. Certainly we would not think their job had been well done if we discovered that their criterion was that the most articulately and forcefully marshalled position should determine the shape of public policy. The public-policy process requires a more rational criterion than that. The basis for developing rational criteria can be found in a further distinction. One might distinguish between normative ethics and public-policy ethics.

In normative ethics, one is concerned with advancing a rationally based set of values, norms, and obligations, together with arguments as to why they are appropriate for guiding human behavior, quite independent of whether law or public policy should restrict or discourage any of those behaviors. Public-policy ethics—a term I find less ambiguous than public ethics (which describes all applied ethics)[27]—is that area of normative ethics that is concerned with determining what sort of policy or decision ought to be made concerning a particular matter of public concern in the social order. Public-policy ethics "is limited by the particularity and the pressing nature of [the policy problem being addressed] and by the resistance of social structures."[28] Public-policy ethics responds to the question: Among all the moral values, ethical positions, moral principles, and so on that are articulated and justified in normative ethics, which should this particular public policy protect and according to what set of priorities?

Following this distinction—and for reasons that I mention later—I believe it is justifiable in terms of public-policy ethics not to ban laboratory in vitro fertilization research or clinical in vitro fertilization followed by embryo transfer even though I believe that some of the activities and consequences associated with these practices are wrong in terms of overall normative ethics.

We return, then, to the question: What are the criteria for selecting the moral values and principles that should give shape to public policy for in vitro fertilization and embryo transfer? I have adapted several of the following five criteria from Isidore of Seville's (c. A.D. 560–636) natural law ideal regarding positive law[29]:

1. *Not all conduct that is arguably immoral should be excluded by prohibitive public policy, nor should all (arguably) moral conduct be supported by public policy, but only those that have a serious impact on the public order*. Paul Ramsey follows this criterion, offering the following example: "A person may believe . . . that suicide is morally wrong and yet oppose any law against suicide," for "the death of a person by self-willed destruction," though done "by (believably) immoral means . . . , need not be regarded as a matter of national interest or public-policy concern."[30] This criterion assumes that some values and principles (for example, certain freedoms), some institutions (for example, marriage), and certain character traits (for example, parental devotion to the nurturing of one's children) are arguably necessary for the public order or for the common good and some are not. The problem in applying this to in vitro fertilization and embryo transfer is to determine what is in the public interest or for the public order. The relevant rights and principles, in many instances, have not been rigorously thought through, principally because there is little precedent in public policy governing these actions. For example, is there a right to advance knowledge about reproductive biology through research? If so, is it in the public interest or required by the public order

that little, if any, limitation be placed on this research? The answers to these questions should have a strong bearing on in vitro fertilization public policy.

2. *In the formulation of public policy, when policymakers must choose between two evil means or consequences, they should choose the lesser evil.* In this criterion, evil effects anticipated from a public policy can be justified by greater good effects or by the greater reduction of evil effects that the policy would be expected to bring about. For example, can the offense against the dignity of human life involved in the planned discarding of human embryos be justified by the faster pace at which an unrestricted policy would permit scientists to discover means of treating genetic and chromosomal abnormalities? By the criterion of the lesser evil, probably little or no in vitro fertilization research would be permitted if either of the slippery-slope arguments were convincing—that, once permitted, in vitro fertilization would (1) unavoidably lead to, or (2) could without logical contradiction extend to, a host of totally undesirable consequences, for example, culture beyond the blastocyst stage, even leading to full-term ectogenesis (thus avoiding use of the human womb entirely); the formation of hybrids or chimeras, by combining human with human or human with nonhuman genetic materials; nuclear transplantation or cloning; storage and banking of living human embryos, perhaps commercially; the use of surrogate wombs; and so on.[31] My own conviction is that most, if not all, of these consequences can be isolated in the public-policy process and that distinctive sets of principles can be used to restrict or prevent them. From a moral perspective, much depends on how clearly and specifically one formulates the principles that are to govern any phase of embryo culture, embryo research, and embryo transfer. In fact, the report of the Ethics Advisory Board has suggested possible legal objections to a number of these individual undesirable consequences.

3. *Public policy should not restrict the rights of members of a society unless the restriction is truly necessitated by the public good.* The more fundamental the right, the more compelling must be the necessity that infringes the right. In developing in vitro fertilization policy, the difficulty is in knowing what moral rights exist and which moral rights should be legally sanctioned. For example, does everyone, married or not, have a right to procreate, such that he or she should not be hindered from having access to in vitro fertilization if that is the only possible means for achieving procreation? And if such a right exists, could there ever be a public-order reason for restricting federally funded in vitro fertilization research to married couples?

4. *Public policy should not be excessively burdensome for people to fulfill or for public authorities to enforce.* For example, a total ban on all in vitro fertilization research and clinical application would not be feasible

because the policy could not be enforced and because such a total restriction would not be practicable in terms of the values of our society to which public policy is necessarily restricted.

5. *When an impasse is reached in determining what is the greater good or lesser evil, a procedural resolution may be the only workable solution.*[32] The principal current federal regulation on in vitro fertilization and embryo transplant, discussed previously, is an example of a procedural regulation—that is, the requirement that a research proposal for DHHS support be first reviewed by an ethics advisory board.

Ethics can make a distinctive contribution to the policymaking process when ethicists advise politicians and administrators on the application of these five criteria governing the relation of moral principles to public policy. Public-policy ethicists make this contribution by performing the following four tasks.

First, they can articulate the nature of the moral conflict(s) and clarify the applicable moral principles. Second, they can elucidate proposed policy options in light of relevant moral principles and their understanding of the dominant public ethos. Third, they can argue for a particular ranking of moral options for public policy and what the probable outcome might be if one moral principle is ranked over another. Fourth, they can offer a critique of the trend in which the policymaking process is going by offering an alternative set of conceptual analyses and normative theories for a final critique of the policy and to encourage subsequent reexamination of the policy.[33] The fourth task is important because, as Robert Neville has pointed out, the more sensitively and thoroughly a policymaking body listens to all opinions and then articulates and reinforces the dominant ethic—currently a liberal ethic—the more difficult it is to cultivate support for fundamentally different perspectives.[34]

Ethical Issues

In vitro fertilization is a burning question for contemporary public policy largely because it represents a dramatic turning point in the technologizing of human life. We are not simply discussing the joining of gametes in petri dishes but the significance and limits of utilizing laboratory techniques in a human event traditionally regarded as sacred—the creation of new human life and of the intimate personal and social circumstances into which that life will be received. Technology, including reproductive technology, can bring benefits, but it would be foolish to do everything modern science and technology make possible simply because they can be done. Human values and goods transcend the technological; these values might prompt us to use reproductive technologies but also to restrict their use.

In the remaining sections of this chapter, I focus on both general normative ethics and public-policy ethics as they bear on in vitro fertilization and embryo transfer.[35] Specifically, I single out some of the dominant moral conflicts that have been presented to policymakers, articulate relevant moral principles, and argue for a particular ranking of moral options in the area of in vitro fertilization and embryo transfer. I do this, however, with the realization that the task of analyzing and applying moral principles is very much conditioned by the broader philosophies and ideologies that govern the development and use of various technologies. The following discussion of specific moral problems and principles should be seen as an attempt to examine at least some of those broader assumptions.

Separating Parenthood from the Sexual Act

The basic issue in both normative and public-policy ethics is that of parenthood and its relationship to human sexuality and procreation. Unfortunately, we have not engaged in sufficient reflection on that issue. The parenthood-related issue arises most clearly in the writings of Paul Ramsey, who argues that in vitro fertilization is wrong because it "pays disrespect to the nature of human parenthood."[36]

The starting point of Ramsey's argument is the act of sexual intercourse that tends, of its own nature, to the natural ends of the strengthening of love (the unitive good) and toward the engendering of children (the procreative good). In addition to nature, Ramsey adds a theological (scriptural) source of knowledge, which teaches that God created the world because of his love and created nothing apart from his love.[37] Therefore, he concludes that the procreative and unitive aspects of sexuality belong together.

On this basis, Ramsey formulates the principle that an ethic that in theory sunders the sphere of personal love from the sphere of procreation is wrong. He concludes that contraception and voluntary sterilization for eugenic purposes—even to the point of excluding procreation for a lifetime when done for a sufficient reason—are permissible, for they separate only "the sex *act* as an act of love from whatever tendency there may be in the act . . . toward the engendering of a child."[38] He reaches a very different conclusion regarding in vitro fertilization—namely, that it is wrong because to procreate by this means is "to procreate from beyond the sphere of love."[39]

One of the principal difficulties with the position espoused by Ramsey (and similar views held by others[40]) is that, by linking the nature and responsibilities of parenthood so closely with the sexual procreative act,[41] he offers a questionable explanation of the meaning of parenthood and consequently unnecessarily conflates the principle of responsible procreation with the principle of responsible parenthood. I would like to offer separate comments on these two principles by beginning with an analysis of the meaning of parenthood and of the principle of responsible parenthood.

Admittedly, the act of sexual intercourse frequently has the capacity to cause conception, thereby making an individual a progenitor though not necessarily a parent. Parenthood is a kinship type of relationship that may or may not follow one's own procreative activity and that may indeed follow upon someone else's procreative activity. Yet parenthood involves more than just procreation; it entails entering into a relationship and assuming a responsibility for the rearing of offspring regardless of its origins. Hence my thesis is that parenthood is entered into by an election or choice and is maintained by a distinctive kind of caring commitment.

One verification of this thesis can be constructed by reflecting on the powerful scriptural image of adoption. In the Epistle to the Romans, Paul says that we become the children of God and know that we are the children of God through the spirit of adoption, whereby we call God our Father.[42] In trying to clarify the Christians' spiritual relationship with God by analogy with an adoptive child's relationship to its parent(s), Paul provides us with imagery that clarifies the meaning of human parenthood generally.

This analogical source of knowledge reveals the following characteristics of parenthood:

1. Parenthood comes about by election.[43]
2. It does not depend essentially on congenital roots.[44]
3. A distinctive spirit (a disposition or mentality),[45] called the spirit of adoption, causes the vital spiritual bond that constitutes the parent-child relationship.
4. This same spirit enables the child to know his or her status of child and to have a radical confidence in the parent.
5. To be a child means to have the right of inheritance—that is, to become the owner and ruler of his or her parents' estate.

The foregoing account of the meaning of parenthood has not been intended as an appeal to scripture as moral authority for establishing the "correct" Christian or Judaeo-Christian view of parenthood; rather, it has simply been a philosophical attempt to clarify the meaning of parenthood through reflection on the theological analogy suggested by Paul. A second way of verifying my thesis is by direct reflection on the meaning of contemporary adoptive parenthood—a move that is prompted by the Pauline text, but which is also appropriate in terms of in vitro fertilization because in vitro fertilization, like adoptive parenthood, depends on procreation's occurring outside the parents' own bodies.

When one reflects on the reality of parenthood in the case of adoptive parenthood, one is forced to examine parenthood with a focus and intensity that is not so necessitated when parenthood is initiated through (one's own) sexual activity, which of its nature is filled with many ambiguities. The opportunity to learn more about parenthood and other human relations by reflection on the realities of adoptive parenthood is a relatively new opportunity. Although adoption has existed since ancient times for limited purposes—for example, to facilitate inheritance procedures—it is apparently only recently that adoptive parenthood has become a kinship tie equivalent—psychologically, emotionally, socially, and legally—to parenthood initiated with congenital ties.[46] As with parenthood initiated by one's own procreative activities, adoptive parenthood consists of the choice and commitment to nurture, love, and care for this new life in a parental role. This may occur prior to, simultaneously with, or subsequent to the conception or the arrival of the child. Hence, in reply to Ramsey, parenthood does not, as a conceptual matter, find its core meaning in the customary means of procreation—that is, sexual intercourse.

Yet it is often said that the purpose of in vitro fertilization is to make it possible for an infertile woman or couple to become parents by having their own child. They can, however, have their own child through adoption. Does birth parenthood, then, represent no further values beyond those that can be experienced through adoptive parenthood? Yes, I would suggest that it adds four things that people value: (1) it proves that one has a reproductive power that can be activated; (2) it provides an experience that puts one in touch with biological reproduction in a personal way; (3) it makes a baby that may have physical features like the progenitors'; and (4) it continues a genetic lineage that can be an object of pride and respect. I believe the element of lineage is the most significant. The apparently deep-seated desire on the part of many adopted children to discover their biological roots seems to be indication of the human value of genetic lineage. In the final analysis, one can say that the consanguinity element found in being the progenitor of one's children represents a distinctive value and motivation for parenthood, but it is not essential to the very meaning of parenthood; and when one assesses what is required for public order, preserving one's personal lineage represents a value that should be ranked lower than other more important values—for example, respect for individual living human beings and maintaining the health of the living.

On the basis of this conceptual analysis of parenthood, we can articulate the moral principle of responsible parenthood. The responsibility associated with the parental role is based on the needs of the child—the child's prolonged need for care, affection, and protection, not only for survival but also for the experiences essential to his or her total development.[47] Parental responsibility requires one to respond to those needs by committing oneself to a parent-child covenant and by pursuing parental virtues—to care for, to love, and to loyally pursue the welfare of one's children within a permanent family group.

How does this normative concept of parenthood relate to the task of public policymaking in the area of in vitro fertilization and embryo transfer? Public policy cannot decide what parenthood means, but it can determine what responsible parenthood requires in terms of the public good. I would conclude that in vitro fertilization and embryo transfer, used clinically as a means of becoming parents, does not hinder the development of true parenthood (by election and commitment), nor does it, in a manner that would be harmful to the public good, place an obstacle in the way of fulfilling the requirements of the principle of responsible parenthood.

Does this mean that any sort of reproductive intervention—not only in vitro fertilization but also the use of surrogate wombs, artificial insemination using donor's sperm, cloning of humans, and others—would be permissible because producing humans by such techniques would not offend against the principle of responsible parenthood? No, not at all, for there could be moral objections to those actions in public policy if they offend against the principle of responsible procreation, a question that is separable from that of responsible parenthood, as the preceding analysis shows.

Separating Procreation from Marriage

Thus, public policymakers are posed with two distinct moral questions if they are to take seriously objections of the type raised by Paul Ramsey: (1) Should parenthood be separated from the sexual act? and (2) Should procreation be separated from marriage? We now reflect on the second of these two questions.

My thesis is that, in terms of public-policy ethics, the preferred moral option is that governmentally sponsored in vitro fertilization with embryo transfer be restricted to the marital context—that is, that it be restricted to married couples who are attempting to bear their own child by use of their own gametes.

At the level of general normative ethics, one reason why the procreation of new human life ought to be intimately joined with marriage is because that covenant of love (marriage) provides essential means and assurances for parenting and for the nurturing of the child. I agree with Ramsey that human procreation should not be subjected merely to the requirements of technical control divorced from a loving atmosphere and the covenants that engender and protect that atmosphere.[48] For the same basic reason, the husband and wife should use their own gametes for the procreation of the child. The use of donor's ovum or sperm and the use of a host womb are ethically objectionable on the grounds that these procedures, in principle, are not consonant with the sexual-physical-relational exclusivity and unity of the marriage convenant. This exclusivity and unity make marriage a source of the stability that responsible parenthood requires.

In terms of public-policy ethics, it is entirely reasonable for public policy to restrict publicly funded in vitro fertilization and embryo transfer to married couples because it is in the public interest to provide some general assurances that human procreation will be followed by responsible parenthood; and as long as there is the assumption that marriage and the family provide children with care, love, loyalty, and support with a better assurance than any other institution, the government has an interest in making these assurances. One respected legal source has suggested that the government interest in "fostering marriage and discouraging illegitimacy" might be proffered as a rationale in support of a law restricting publicly funded in vitro fertilization and embryo transfer to married couples.[49]

What reason can be given in terms of public-policy ethics for restriction of government-supported in vitro fertilization research to the gametes and womb of the subject couple? In addition to the principled reason given before, one can add that public policymakers should be concerned about avoiding the psychosocial harm that would come from humans' being begotten from randomly obtained and possibly anonymous sources. If a governmental agency were to sanction the use of extramarital gametes, this would communicate the message to the community that it does not make any difference whose genetic materials are used to beget a child—and that policy would trivialize human procreation. Thus, there are good, socially sanctioned reasons why government-supported in vitro fertilization or in vitro fertilization research should remain within the personal and exclusive union of married couples.

One objection against making marriage a prerequisite to government-supported in vitro fertilization would be the existence of a right to in vitro fertilization on the part of everyone, including the unmarried, unless, of course, a sufficiently strong reason can be given for the marriage requirement. For example, is there a right to procreate? If so, individuals may have a right of access to in vitro fertilization, if that is the only available means of achieving procreation. In the law, it appears much clearer that there is a right to privacy—for example, a right of bodily privacy that excludes compulsory sterilization,[50] or a right to marital privacy not to be compelled to forgo use of contraceptives.[51] It does not seem philosophically convincing to claim that there is a fundamental right to procreate for there is no convincing public-interest reason why the government or other public agencies should be obliged to accomplish the fulfillment of every person's procreative faculties. Yet there is a negative right, both morally and legally, protecting individuals from unwarranted intrusion into personal and family life in the procreative sphere. However, as the Ethics Advisory Board concluded, one could not claim that restriction of access to in vitro fertilization is interference with the right of marital privacy and with a married couple's right to choose whether, and in what manner, to achieve procreation.[52] Therefore, there does not appear to be sufficient basis in a rights argument against the marriage-requirement thesis offered here.

Another objection to this thesis would approach from the opposite direction. This objection might claim that my thesis does not go far enough. Procreation should take place not just in marriage but through marital intercourse. In vitro fertilization would be morally excluded because, as Ramsey claims, by taking place in a laboratory instead of in the intimacy of the sexual act, it sunders the sphere of personal love from the sphere of procreation.

I would offer two reasons in disagreement with Ramsey and others who offer similar arguments. First, it is inadequate to identify the in-principle unethical character of an act or the intrinsic evil of a kind of action in terms of the physical structure of the (sexual) act without taking more directly into account the (procreative and parental) values, intentions, and responsibilities surrounding that type of action. Second, even granting Ramsey's general principle, in vitro fertilization is not an *in-principle* separation of procreation from the sphere of love. A loving couple, desiring to procreate as an expression of their love and as the fruit of their sexualilty, and further intending to raise their child in the circle of their love, do not sunder the procreative and loving dimensions of their sexuality by joining their gametes outside their bodies and by returning the fertilized egg within a few days. Aside from this unobjectionable set of personal intentions, nothing in the in vitro fertilization process itself in principle signifies the sundering of procreation from marital love.

Nonetheless, I would agree with Ramsey to the extent that the principle of responsible procreation requires a loving and caring atmosphere before one brings a human being into existence; but that requirement is provided, in principle, by restricting in vitro fertilization procreation to married couples.

Among all the ethical issues in in vitro fertilization and embryo transplant, I have deliberately given most attention to the first two—interlocking problems raised by the roles of parenthood and procreation—because I believe they are fundamental issues not sufficiently examined in the literature. For the remaining issues, I must be unsatisfyingly brief, but my purpose is the same—namely, to show how reflection on moral principles relates to the requirements of public policy.

Destruction and Loss of the Embryo

Destruction or loss of the human blastocyst or embryo can occur for different reasons in in vitro fertilization. For the sake of efficiency, several ova may be extracted and fertilized, and those not selected for implantation will be discarded. Other embryos may be discarded because they show signs of defect. Still others may simply be lost because they cannot survive in culture or in the womb following embryo transfer.

Is it a wrongful moral act to cause the destruction, abandonment, or loss of a human embryo? The answer to this question depends in part on the answers to the questions: What is the moral status of the human embryo? Must human life be protected—for example, in a research setting—at this early stage of development?

Questions of this kind have been thoroughly discussed in the debates on abortion and in the controversies surrounding research on living but aborted fetuses. Yet the discussion in abortion is quite different from that in the in vitro fertilization debate. Frequently, arguments favoring abortion have not been based on the belief that the human fetus has no moral value but on the principle that in a conflict situation the interests, wishes, or alleged rights of the pregnant woman should prevail. However, as Leon Kass has pointed out, the fate of these embryos is not in conflict with the claims, wishes, or alleged rights of a pregnant woman.[53] Thus, the question here is different: Should we deliberately produce human embryos in vitro for the express, sole purpose of laboratory research?

In terms of general normative ethics, I would approach the question this way. After fertilization is complete, there exists a new and distinct human entity, with its unique genetic identity—a live, organic whole that is "fully potent for the self-initiated development into a mature human being, if circumstances are cooperative." From this point on, there is "continuity of subsequent development, even if the locus of the embryo alters with implantation (or birth)." The blastocyst itself is not fully a human being; it is not yet organized into distinctive organs, cannot think, reason, or relate socially—stages that will occur only later in pregnancy and after birth. However, it is a human life; it "is not humanly nothing; it possesses a power to become what everyone will agree is a human being," provided it is implanted in a woman. And the blastocyst "elicits, or ought to elicit, our feelings of awe and respect" for several reasons; because of its origins, because of what it is developmentally, and because it has the potential to be a mature human being if all goes well.[54]

This early human life in vitro commands respect, but what sort of respect, and how should that respect influence our behavior? I would say it requires at least that we avoid doing it harm, partly because that is the most basic principle for protection of the human community, and partly based on the principle that when there are doubts about the claims that a human life makes on us, we must give that human life the benefit. The application of this norm to the blastocyst is also called for by the consideration that there is no later dividing line in the development of the human before which there is not and after which there is the basis for this minimal protection. Legally, of course, the individual acquires the full range of rights as a person after birth.

An application of these principles of general normative ethics would mean that fertilized ova through the blastocyst stage of development should not be directly harmed by the administration of toxic drugs or noxious stimuli even for a worthwhile research purpose. Furthermore, both the respect we owe the blastocyst (minimally, that of doing it no harm) and the principle of responsible procreation require that we not produce a human blastocyst with the intention of abandoning or extinguishing it.

Thus far I have been discussing the moral status of the human embryo in terms of general normative ethics. To what extent should we expect these principles to be incorporated in public policy? Public-policy ethics requires us to take into account the principle of the lesser of two evils in a context in which the possibility exists of no principled restraint against the making of embryos, and eventually babies, in what might be the most dehumanizing of conditions, in both the short-term and the long-term assessment. Yet the research is being pursued by scientists who want to accomplish significant and desirable human benefits—developing or testing more-adequate contraceptives, determining causes of infertility, investigating circumstances leading to malignant tumors, studying mechanisms by which chromosomal abnormalities are produced, and increasing understanding of normal and abnormal cell growth and differentiation.[55]

A resolution of this dilemma, in my view, hinges on the distinction that can be made between clinical research and basic laboratory research. Clinical research focuses on a particular problem manifested by a patient or patients, with an attempt to understand and/or ameliorate the problem. In this case, the problem is infertility. Nontherapeutic clincial research emphasizes only the understanding of some problem, while therapeutic research includes its amelioration as well. Laboratory research is done in and for the sake of more complete knowledge and understanding of some issue—in the case at hand, to increase human knowledge of human genetics, particularly of the fertilization and developmental process, but not in reference to a specific subject couple or to securing a safe delivery of a healthy baby following in vitro fertilization.

Making use of this distinction, I believe policymakers should consider the following ranking of moral options for inclusion in public policy. These proposed public-policy rules also afford an opportunity to summarize my principal public-policy conclusions.

1. Laboratory research on human in vitro fertilization and development should not be conducted or supported by a governmental agency unless it is in the context of clinical research. The mystery of human growth and development can easily be trivialized into a technical problem of management, in which the respect that is due to every developing human life can easily be overlooked. For this reason I would recommend that studies on in vitro fertilization and embryo culture and transfer always be conducted within the context of clinical research. Thus, in disagreement with Richard McCormick and the Ethics Advisory Board, I would exclude, on the basis of the principle of respect for human life, any studies that intend the fertilization and destruction of embryos.[56] When one considers the prospect of one particular proposal that envisioned fertilization of 450 human ova, and when one considers the likelihood of this project's being multiplied many times over, one is struck by the depersonalizing nature and ultimately the desensitizing effect of such research activity.[57] For these moral reasons, research should be restricted to the context of clinical research involving an infertile married couple where the human and moral significance of this work cannot so easily be overlooked, even if this means that more steps may have to be taken than if the work were done in the laboratory-research setting.

2. Therapeutic clinical research in in vitro fertilization, embryo culture, and embryo transfer may be conducted provided an institutional review board has determined that:

a. The research is scientifically sound and significant;
b. Adequate prior studies on animals have been conducted;
c. Adequate provisions have been made to acquire informed consent from the prospective parents and to protect the privacy of parents and of children conceived by this process;
d. Only married partners who are presumed infertile are accepted as subjects of these studies, and only ova provided by the wife and sperm provided by the husband are accepted for use;
e. When, in spite of the general intention to implant the embryo, the decision is made that the embryo cannot or even ought not be sustained in reference to achieving a pregnancy in the subject couple, efforts to sustain the embryo should not be continued beyond the stage when implantation would normally be expected to occur—that is, within approximately fourteen days after fertilization;
f. Termination of the pregnancy may not be imposed on the subjects by the investigator if, for example, the investigator discovers defect in the embryo. When medically indicated, abortion will be permitted only on condition that informed consent is obtained from both parents.

3. Nontherapeutic clinical research in in vitro fertilization, embryo culture, and embryo transfer should not be conducted unless an institutional review board has determined that:

a. All the provisions of recommendation 2 have been followed,
b. Fertilized ova through the blastocyst stage of development are not directly harmed by the administration of toxic drugs or noxious stimuli.

Some of the stipulations in recommendations 2 and 3 require little or no explanation: the scientific reliability and importance of the protocol itself, prior studies in animals, informed consent, and refraining from coercing patients to abort. Recommendation 2d, restricting the research to married couples, was explained previously in terms of the principle of responsible procreation. Recommendation 3b, on not exposing the embryo to excessive harm, has already been explained in terms of the principle of respect for human life. Recommendation 2e, requiring a fourteen-day limit on research, requires further justification.

U.S. policy should take seriously into account, without necessarily being absolutely bound by, the sentiments of a large segment of its population—that life at the earliest stages should be respected and protected. Yet within the one moral and religious group that articulates this position so strongly—the Roman Catholic community—there is considerable doubt among theologians, philosophers, and physicians about the moral status of the preimplantation embryo. During those first two weeks or so, while there is still possibility of twinning and the formation of polyploid chimeras, some ethicists claim there is not present the individuality required to establish the moral claims of a new, truly human individual. Put in other words, the material or biological uniqueness usually associated in the scholastic tradition with ensoulment or personhood is not clearly present. Consequently, even within this tradition, an objective rational doubt exists as to the moral status of the blastocyst. Although I do not agree with those who hold that a new metaphysical status of truly human life occurs after about two or three weeks of life, and that prior to then these lives are "not deserving of the protection of truly human life,"[58] I believe that that position among Roman Catholic scholars offers sufficient reason, on extrinsic or public-policy grounds, for acknowledging the fourteen-day limit in public policy. It should also be noted that the fourteen-day point is where the DHEW guidelines on fetal experimentation begin their protection of the fetus—that is, with implantation.

Would allowing the embryo to perish after approximately fourteen days, under certain conditions, be opposed to the principle of respect for human life? Support for a negative reply to this question can be found in the high rate of human-embryo loss following in vivo fertilization. In a report on intrauterine mortality, Dr. James Schlesselman indicated that out of 100 ova exposed to the risk of fertilization, of which 84 are fertilized, only 37.2 percent survive a live birth. Therefore, 63 percent die, and approximately 44 percent die in the first two weeks after fertilization.[59] One should not too quickly conclude from this that we are permitted to mimic nature in this regard. If we want to argue from these data that the embryo has a reduced moral status because of the instability of its life, what are the metaphysical implications of such a conclusion? What sort of stability is required, at the beginning and end of life, to establish that there is human life that calls for respect and protection? Furthermore, at the level of ethics, not every result produced by unassisted nature may by that fact be mimicked in deliberate human behavior: to claim that position would be to commit the so-called naturalistic fallacy. Yet we are clearly not obliged to avoid every risk of defect or of early death, for that would be to require too much by way of physical perfection. I believe that there is a way of resolving this dilemma, not at the level of metaphysics or ethics but in terms of a reasonable public-policy assumption. For public-policy purposes, the customary human willingness to accept the risk of loss of embryonic life in vivo argues for the public acceptability of approximately the same level of risk of loss of embryonic life in laboratory research, especially when there is biological indication that the embryo has a defect that would most likely make it impossible for the individual to survive in a normal uterine environment.

In an earlier statement submitted to the Ethics Advisory Board, I acknowledged—on the basis of public-policy ethics—the acceptability of the fourteen-day moral option.[60] The Ethics Advisory Board later adopted the fourteen-day standard for research in this area but differed in its conclusions from the aforementioned recommendations, principally in permitting laboratory in vitro fertilization research—although even here the board gave a clinical orientation to the research by stipulating that the research must be designed primarily "to establish the safety and efficacy of the embryo transfer" when that information is "not reasonably attainable by other means."[61]

Risk of Defect to the Human Embryo

Paul Ramsey opposed federal policy supporting in vitro fertilization because of "the *irremovable* possibility that this manner of human genesis may produce a damaged child."[62] That would be wrong, he claimed, because to put our progeny at risk of unknown and unknowable damage from the very procedure by which they are procreated violates the primary principle of medical ethics: Do no harm. This view is not unique to Ramsey. In 1971, Leon Kass argued against in vitro fertilization, claiming it involves choosing unknown risks of damage to the child, coupled with the impossibility of obtaining the unborn or unconceived child's informed consent for nontherapeutic research (that is, not for the benefit of the developing embryo).[63] In 1979, Kass changed his position on this.[64]

Should this ethical issue lead us to revise any of the policy recommendations given earlier? No, Ramsey's view would not seem to be convincing for it would appear to rule out the ethical acceptability of procreation through mating. "Natural" human procreation and gestation involve both known and unknown risks of defect to the humans thus begotten. As Mastroianni has pointed out, "Those who favor human embryo transfer concede that it is impossible to guarantee the genetic normality of in vitro–handled material but suggest that it is equally true that each normally produced conceptus has no guarantee of future health."[65] Schlesselman described to the Ethics Advisory Board nature's way of handling at least some genetic abnormalities through spontaneous abortion in the following terms. If one considers the relationship of chromosomal anomalies to spontaneous abortion, 66 percent of spontaneous abortions in the first seven weeks of development are affected with abnormal karyotype, and a total of 81 percent of the fetuses that die before birth are chromosomally abnormal.[66] To do a proper risk assessment, further information is needed on the extent to which the in vitro fertilization and embryo-transfer process introduces risk of defect, data on in vivo–conceived early embryonic life, information on defects appearing at a later age among those conceived by in vitro fertilization, the extent to which genetic or chromosomal defects possibly introduced by in vitro fertilization might be screened out by the natural process by which most such embryos are lost early in gestation, and so on.

From the perspective of public-policy ethics, the customary and apparently reasonable human willingness to accept the risk of defect to new embryonic and fetal life in vivo argues for the public acceptability of approximately the same level of risk of defect to embryonic and fetal life in in vitro fertilization and in the subsequent pregnancy.

The Problem of Just Allocation

Should the government actually fund research in in vitro fertilization, embryo culture, and embryo transfer, as well as the clinical application of these techniques to couples with an obstructed oviduct? A response to this question entails application of the principle of the just distribution of benefits and burdens in society. There are several reasons why, in my view, these activities should not be given a high priority for funding by government agencies:

1. More basic health needs exist in society, especially in the area of preventive medicine;
2. The satisfaction that comes from knowing we are survived by those who are genetically like us is not a greater public good than the more-pressing health needs in our global society;
3. Funds are available for these research and clinical activities from private sources;
4. Public funding of these activities, especially of research on embryos without subsequent implantation, would cause a traumatic division in this country already deeply divided on the question of the value of prenatal human life.[67]

Conclusion

This chapter has dealt with the relationship of ethics to public-policy (federal) formation in the area of human in vitro fertilization and embryo transfer in both the research and clinical settings. Three historical events had a significant influence on policymaking in this area: (1) the dramatic and sensationalized breakthrough in 1978 of the world's first test-tube baby, which had not been preceded by widespread public discussion of the ethical issues; (2) just one month later, the Del Zio court case, involving the destruction of a fertilized human ovum without the mother's consent; and (3) the 1973 U.S. Supreme Court *Roe* v. *Wade* abortion decision, which tilted the balance in favor of the woman's right to privacy—a decision that accounted for opposition in the U.S. Congress to fetal research, eventually resulting in a federal moratorium on in vitro fertilization research. This article also reviewed federal policy on in vitro fertilization research and embryo transfer, noting in particular the provision that no DHEW funds can be used for this purpose until the proposal has been reviewed by an ethics advisory board—a ruling that still stands.

So often, discussions of the ethical issues related to in vitro fertilization take place in a vacuum insulated from the pressures of the public-policy forum. This chapter tried to bridge this gap by offering criteria for determining how moral principles should selectively shape public policies and by explaining the tasks of both general normative ethics and of public-policy ethics in applying these criteria and assisting the policymaking process.

Five ethical issues were discussed in the context of public-policy ethics. First, is in vitro fertilization ethically objectionable because it separates parenthood from sexual intercourse? Because this traditional objection conflates procreation with parenthood, I separated the problem of responsible parenthood from that of responsible procreation, offered a close analysis of the meaning and moral requirements of parenthood, and concluded that in vitro fertilization and embryo transfer, used clinically as a means of becoming parents, does not offend against the principle of responsible parenthood. Second, I concluded that the principle of responsible procreation would indicate that governmentally sponsored in vitro fertilization with embryo transfer should be restricted to married couples who are attempting to bear a child by use of their own gametes—thus providing general assurance of a stable and committed atmosphere for fulfilling the parental responsibility of nurturing the child. Third, destruction and loss of the human embryo in in vitro fertilization and embryo-transfer research raises questions about the moral status of the embryo. Avoiding language about the rights of the blastocyst, I suggested that this early stage of human life elicits awe and calls for respect, at least to the extent of not harming it. In the public-policy arena, the principles of respect for human life, responsible parenthood, and responsible procreation would support the conclusion that laboratory research on human in vitro fertilization and development, intending the fertilization and destruction of human embryos, should not be conducted or supported by a governmental agency. However, therapeutic clinical research would be allowed, within the customary research restrictions (such as informed consent), for married partners, utilizing their own ova and sperm, provided the embryo is either transfered to the uterus or, for sufficient reason, is allowed to perish within fourteen days. Nontherapeutic clinical research would be permissible, with the added proviso that the blastocyst not be directly harmed by the administration of toxic drugs or noxious stimuli. I gave public-policy reasons for these conclusions and compared them with the conclusions of the DHEW Ethics Advisory Board.

The fourth ethical issue was the risk of defect to the human embryo. Similar to the issue of the risk of loss of the life of the fertilized ovum, the resolution of this question revolved about the parallel reasonableness of accepting risk of defect to new embryonic and fetal life in vivo. The problem of just allocation of resources was the fifth and final issue discussed. Four reasons were given why public policy should not give high priority to the funding of in vitro fertilization and embryo-transfer research by government agencies.

These conclusions are offered with an awareness of the magnitude of the decisions to be made and of the need for further discussion and debate.

Notes

1. P.C. Steptoe and R.G. Edwards, "Birth after the Reimplantation of a Human Embryo," *Lancet* 2 (1978):366; Don Hilton Hilson; R.L. Bruce; and D.G. Sims, "Successful Pregnancy Following In-Vitro Fertil-

ization,'' *Lancet* 2 (1978):473; P.C. Steptoe and R.G. Edwards, *A Matter of Life: The Study of a Medical Breakthrough* (London; Hutchinson & Co., 1980); and P.C. Steptoe; R.G. Edwards; and J.M. Purdy, ''Clinical Aspects of Pregnancies Established with Cleaving Embryos Grown 'In Vitro','' *British Journal of Obstetrics and Gynaecology* 87 (1980):757–768.

2. Mark I. Evans; Anil B. Mukherjee; and Joseph D. Schulman, ''Human In Vitro Fertilization,'' *Obstetrical and Gynecological Surgery* 35 (1980):71–81.

3. George F. Will, ''Irreverent Test Tubes,'' *Washington Post,* 30 July 1978, p. B7.

4. George F. Will, ''Man-Made Evolution,'' *Washington Post,* 3 August 1978, p. A23.

5. Partial transcript, charge of court and verdict of jury, authorized facsimile, Del Zio v. Presbyterian Hospital (S.D.N.Y. 17–18 August 1978), pp. 4370–4475, no docket no. available.

6. Mark I. Evans and Alan O. Dixler, ''Human In Vitro Fertilization: Some Legal Issues,'' *Journal of the American Medical Association* 245 (1981):2326.

7. ''Fetal Research: Pulling the Stopper,'' *Medical World News,* 13 September 1974, p. 4; ''Test-Tube Bereavement,'' *Newsweek,* 31 July 1978, p. 70; and ''Woman is Awarded $50,000 in Test-Tube Embryo Case,'' *Washington Post,* 19 August 1978, p. A4.

8. Roe v. Wade, 410 U.S. 113, 35 L. Ed. 2d 147, 93 S. Ct. 705 (1973).

9. Ibid., p. 161.

10. Ibid., pp. 163–164.

11. Victor Cohn, ''Live Fetus Research Debated,'' *Washington Post,* 10 April 1973, pp. A1.

12. U.S., Congress, House, ''A Bill to Prohibit the Use of Appointed Funds to Carry Out or Assist Research on Living Human Fetuses,'' H.R. 7850, 93rd Cong., 1st Sess., 15 May 1973.

13. U.S., Department of Health, Education, and Welfare, National Institutes of Health, ''Protection of Human Subjects,'' *Federal Register* 38 (1973):31743.

14. U.S., Cóngress, House, *Congressional Record,* 93rd Cong., 1st Sess., 1973, 119, pt. 14:17465–17480; and U.S., Congress, Senate, *Congressional Record,* 93rd Cong., 1st Sess., 1973, 119, pt. 23:29213–29232.

15. U.S., ''National Research Act,'' Pub. L. No. 93-348, 88 Stat. 342 (approved 12 July 1974).

16. U.S., Department of Health, Education, and Welfare, Office of the

Secretary, "Protection of Human Subjects: Proposed Policy," *Federal Register* 39 (1974):30650.

17. U.S., Department of Health, Education, and Welfare, National Commission for the Protection of Human Subjects of Biomedical and Behavioral Research, *Research on the Fetus: Report and Recommendations* and *Appendix,* 2 vols. (Washington, D.C., 1975).

18. Ibid., p. 62.

19. U.S., Department of Health, Education, and Welfare, Office of the Secretary, "Protection of Human Subjects: Fetuses, Pregnant Women, and In Vitro Fertilization," *Federal Register* 40 (1975):33527.

20. U.S., Department of Health, Education, and Welfare, Ethics Advisory Board, *HEW Support of Research Involving Human In Vitro Fertilization and Embryo Transfer: Report and Conclusions* and *Appendix,* 2 vols. (Washington, D.C., 1979). Before making any policy decision, the department published the text of *Report and Conclusions* in the *Federal Register* 44 (1979):35033–35058.

21. Frank H. Marsh and Donnie J. Self, "In Vitro Fertilization at the Norfolk Clinic," *Hastings Center Report* 10 (June 1980):5–6.

22. Ethics Advisory Board, *Federal Register* 44 (1979):35057.

23. This problem is discussed very well in Albert R. Jonsen and Lewis H. Butler, "Public Ethics and Policy Making," *Hastings Center Report* 5 (August 1975):19–31.

24. Vernon J. Bourke, *History of Ethics* (New York: Doubleday, 1968), p. 8, cited by Jonsen and Butler, "Public Ethics," p. 22.

25. David Easton, *A Sytems Analysis of Political Life* (New York: John Wiley & Sons, 1965), p. 358, cited by Jonsen and Butler, "Public Ethics," p. 21.

26. Jonsen and Butler, "Public Ethics," p. 20.

27. *Public ethics* is the term employed by Jonsen and Butler, "Public Ethics," p. 22.

28. Ibid.

29. See M.P. Golding, "History of Philosophy of Law," in *Encyclopedia of Philosophy,* 8 vols., edited by Paul Edwards (New York: Free Press, Macmillan Co, 1967), 4:257. His criteria were repeated in a variety of later adaptations by Gratian, Aquinas, and many theologicans of our times—for example, John Courtney Murray, *We Hold These Truths* (New York: Sheed & Ward, 1960), pp. 166–167.

30. Paul Ramsey, "Testimony on *In Vitro* Fertilization," in Ethics Adivsory Board, *In Vitro Fertilization: Appendix,* essay no. 7, pp. 1–31.

31. The slippery-slope, or wedge, argument is discussed by Leon R. Kass, "Ethical Issues in Human *In Vitro* Fertilization, Embryo Culture and

Research, and Embryo Transfer,'' in Ethics Advisory Board, *In Vitro Fertilization: Appendix,* essay no. 2, pp. 1–37, at pp. 17–20. See also Sissela Bok, ''The Leading Edge of the Wedge,'' *Hastings Center Report* 1 (December 1971):9–11; and Paul Ramsey, ''The Wedge: Not So Simple,'' *Hastings Center Report* 1 (December 1971):11–12.

32. For the inclusion of this principle among these criteria, I am indebted to Richard A. McCormick.

33. The first three tasks are articulated in a slightly different form in Jonsen and Butler, ''Public Ethics,'' pp. 22–24.

34. Robert Neville, ''On the National Commission: A Puritan Critique of Consensus Ethics,'' *Hastings Center Report* 9 (April 1979):22–27.

35. A thorough review of the ethics literature published through 1978 is found in LeRoy Walters, ''Human In Vitro Fertilization: A Review of the Ethical Literature,'' *Hastings Center Report* 9 (August 1979):23–43. This same paper was previously published in slightly different form as ''Ethical Issues in Human *In Vitro* Fertilization and Research Involving Early Human Embryos,'' in Ethics Advisory Board, *In Vitro Fertilization: Appendix,*essay no. 1, pp. 1–52, pp. (FN-1)–(FN-4). Another survey of the ethical issues is found in Ethics Advisory Board, *Federal Register,* 44(1979): pp. 35036–35047, which summarizes the technical and ethical issues as presented to the board in papers prepared by its consultants. Prior to these surveys, the principal multidisciplinary works on in vitro fertilization were G.E.W. Wolstenholme and David W. Fitzsimmons, eds., *Law and Ethics of AID and Embryo Transfer,* Ciba Foundation Symposium, n.s., no. 17 (Amsterdam: Associated Scientific Publishers, 1973); National Academy of Sciences, National Research Council, Assembly of Behavioral and Social Sciences, Committee on the Life Sciences and Social Policy, *Assessing Biomedical Technologies: An Inquiry into the Nature of the Process* (Washington, D.C., 1975), pp. 13–31; and ''Reproductive Technologies,'' in *Encyclopedia of Bioethics,* 4 vols., edited by Warren T. Reich, pp. 1439–1470 (New York: Free Press, Macmillan Co., 1978), vol. 4.

36. Paul Ramsey, *Fabricated Man: The Ethics of Genetic Control* (New Haven: Yale University Press, 1970), p. 33.

37. Ibid., p. 38, referring to John 1.

38. Ibid., p. 34, his emphasis.

39. Ibid., p. 39.

40. An explanation and assessment of the Roman Catholic version of this position and its relevance to in vitro fertilization is presented by Charles E. Curran, ''*In Vitro* Fertilization and Embryo Transfer: From a Perspective of Moral Theology,'' in Ethics Advisory Board, *In Vitro Fertilization: Appendix,* essay no. 4, pp. 1–33.

41. See especially Ramsey, *Fabricated Man,* p. 87.

42. Rom. 8:15.

43. The theme of election or choice is not unique to Paul. In the Hebrew Scriptures, Israel is the Lord's son because of its being chosen by God; see Exod. 4:22; Jer. 31:9; and Hos. 11:1.

44. "It is not the children of the flesh who are the children of God; it is the children of the promise who are to be considered descendants." See Rom. 9:3. This underlines God's choice or election as the basis of the relationship.

45. Because Paul plays on the term *pneuma* (spirit) as meaning either a human spirit (disposition or mentality) or the Spirit of God, it is hard to say which nuance he intends. Within the framework of analogical knowledge that I am using, it is, in any case, appropriate to draw out the secular meaning. For an exegesis of the text on adoptive sonship, see Raymond E. Brown; Joseph A. Fitzmyer; and Roland E. Murphy, eds., *The Jerome Biblical Commentary* (Englewood Cliffs, N.J.: Prentice-Hall, 1968), pp. 315–316.

46. Eugene A. Weinstein, "Adoption," in *International Encyclopedia of the Social Sciences,* 17 vols., edited by David L. Sills, pp. 96–100 (New York: Free Press, Macmillan Co., 1968), vol. 1.

47. See Sally Provence, "Object Relations in Childhood," in *International Encyclopedia of Psychiatry, Psychology, Psychoanalysis, and Neurology,* 12 vols., edited by Benjamin B. Wolman, pp. 104–108 (New York: Aesculapius Publishers, Van Nostrand Reinhold Co., 1977), vol. 8.

48. Ramsey, *Fabricated Man,* pp. 32–52 and 86–96.

49. Dennis M. Flannery; Carol Drescher Weisman; Alan N. Braverman; and Christopher R. Lipsett, "Legal Issues Concerning *In Vitro* Fertilization," in Ethics Advisory Board, *In Vitro Fertilization: Appendix,* essay no. 18, p. 40.

50. See Skinner v. Oklahoma, 316 U.S. 535, 86 L. Ed. 1655, 62 S. Ct. 1110 (1942).

51. Griswold v. Connecticut, 381 U.S. 479, 14 L. Ed. 2d 510, 85 S. Ct. 1678 (1965).

52. Ethics Advisory Board, *Federal Register,* p. 35048.

53. Leon R. Kass, " 'Making Babies' Revisited," *Public Interest,* no. 54 (Winter 1979): 36. This article is based largely on testimony given before the Ethics Advisory Board, which was published in their *In Vitro Fertilization: Appendix,* essay no. 2, pp. 1–37.

54. Kass, " 'Making Babies' Revisited," pp. 37–38.

55. See Ethics Advisory Board, *Federal Register,* p. 35039.

56. Richard A. McCormick, *How Brave a New World? Dilemmas in Bioethics* (Garden City, N.Y.: Doubleday, 1981), p. 329.

57. Kass, " 'Making Babies' Revisited," p. 33.

58. Curran, "In Vitro Fertilization," p. 17.

59. James J. Schlesselman, "How Does One Assess the Risk of Ab-

normalities from Human *In Vitro* Fertilization?'' in Ethics Advisory Board, *In Vitro Fertilization: Appendix,* essay no. 17, p. 37. Also see Schlesselman's article with the same title in the *American Journal of Obstetrics and Gynecology* 135 (1979):135–148.

60. See James J. McCartney and Warren T. Reich, ''Testimony on In Vitro Fertilization Policy'' Before Ethics Advisory Board, Department of Health, Education, and Welfare, 6 December 1978, Springfield, Va., National Technical Information Service, 1979.

61. Ethics Advisory Board, *Federal Register,* p. 35057.

62. Ramsey, ''Testimony on *In Vitro* Fertilization,'' pp. 12, 18, his emphasis.

63. Leon R. Kass, ''Babies by Means of In Vitro Fertilization: Unethical Experiments on the Unborn?'' *New England Journal of Medicine* 285 (1971):1174–1179.

64. Kass, '' 'Making Babies' Revisited,'' p. 43.

65. Luigi Mastroianni, Jr., ''Reproductive Technologies: IV. In Vitro Fertilization,'' in *Encyclopedia of Bioethics,* 4 vols., edited by Warren T. Reich, p. 1450 (New York: Free Press, Macmillan Co.), vol. 4.

66. Schlesselman, ''How Does One Assess?'' pp. 18–20, 50–51.

67. Compare Ramsey, ''Testimony on *In Vitro* Fertilization,'' pp. 3–6.

8 Aspects of Genetical Ethics

John Ladd

This chapter begins with some general remarks about what I call *genetical ethics*—that is, the systematic inquiry into ethical issues surrounding new developments in genetics, in other words, in genetic engineering. I assume that this subject covers positive and negative eugenics, gene therapy, genetic counseling, genetic screening, and so on.

As a philosopher, my first impression in perusing the literature in genetical ethics is that it is a mess—a miscellany of confused and confusing themes, pompous utterances, mythology, dreams, and fairy tales. Except for strictly scientific treatments, very little of substance is available.[1]

Nay-Sayers and Yea-Sayers

For the most part, we find two prevailing attitudes toward genetic engineering. I hesitate to call them theories or doctrines because neither of them advances a specific set of propositions that would be worth examining critically or arguing against. Following Nietzsche, we may call their adherents nay-sayers or yea-sayers. The nay-sayers tend to be theologians, and the yea-sayers tend to be scientists, but there are theologians and scientists on both sides.[2] Fortunately, there are others, including scientists and philosophers, who do not take sides in this categorical way. In any case, the two parties I have mentioned draw the issues in ideological terms and approach them in a doctrinaire rather than a critical fashion. Each party has its favorite clichés, images, and rhetoric; but they both seek to convince by painting scenarios rather than by giving cogent arguments. Their favorite scenario word is *suppose*.[3]

The nay-sayers are obsessed with the wrongness of playing God. It is, one of them says, dangerous to tamper with 3 billion years of nature's wisdom. Their favorite rhetorical device is the slippery-slope argument (sometimes called the camel's nose)—that is, one thing leads to another. Thus, in vitro fertilization leads to adultery, which leads to divorce, which leads to family breakdown, which leads to juvenile delinquency, which leads to crime, which leads to violence and the dissolution of society. If we practice any type of genetic intervention, we will end up with Nazism (or the Brave New World). One slip and we slide all the way down the slope to perdition!

This argument is about as sound as arguing that saying "hello" leads to shaking hands, which leads to kissing, which leads to embracing, which leads to sleeping together, which leads to sexual intercourse, which leads to having babies, which leads to fatherless children, which leads to abandoned children, which leads to crime, and on and on. Or eating one thing leads to eating something else, and so forth, and so eventually you will explode from overeating! It is obvious that a slippery-slope argument is sound only if it can be shown that one step inescapably leads to the next one, as is sometimes the case with drugs, alcoholism, and other types of addiction. Furthermore, the proclivity to slippery-slope arguments suggests that its users have a rather cynical view of the capacity of human beings to make important moral discriminations and that they think that ordinary people are unable to see the moral difference between, say, curing a defective gene and making Frankenstein monsters.

On closer inspection, it becomes evident that the slippery-slope argument derives whatever plausibility it has from the fact that each step in this passage to disaster is something that is implicitly, if not explicitly, forbidden; thus, the slippery-slope argument is just a cover-up for what is basically an ethics of prohibitions—Thou Shalt Nots, that is, a legalistic ethics. When coupled with a bland assumption of moral authority, we have what I call ethicism, the proponents of which call themselves ethicists. Ethicism assumes, mistakenly I believe, that ethics is a branch of knowledge, a set of authoritative propositions or precepts, with regard to which the ethicists are to be deemed authorities—in somewhat the same way as geneticists, say, are deemed authorities on genetics.

Turning now to the yea-sayers, we find that, instead of warning us not to play God, they upbraid us for playing roulette (lottery or dice). We are told that man should take control over his life, over his progeny, and over what happens to the human race. They hold that anyone who challenges the ethical value of genetic engineering is blocking the advance of knowledge and human freedom.

Underlying this sort of yea-saying is a humanism of progressive self-realization, what Monod calls the ethics of knowledge: the "ethic upon which knowledge is founded, and which by free choice makes knowledge the supreme value . . . [and] prescribes institutions dedicated to the defense, extension, the enrichment of the transcendent kingdom of ideas—a kingdom within man, . . ."[4]

If we take a closer look at the yea-sayers' attitude toward genetic engineering, we see behind their view of the relation of knowledge and ethics the outlines of a vast program of consumerism: science—namely, genetics—displays its wares, and we—the consumers—are free to decide what kind of progeny to buy—for example, a clone, a Cleopatra nose, an Einstein brain, old age without aging, and various and sundry sorts of hybrids including a society of superhumans in which aggression, violence, selfishness, stupidity, and other undesirable traits have been bred out.

Quite apart from any questions about the scientific validity of these suppositions, I would like to suggest that they have the relation of science and ethics exactly backward. Ethics is not simply a matter of shopping around for the good life and choosing, for example, between goods to be gained or evils to be avoided; it does not start with a survey of a multitudinous world of possibilities offered to us for our selection. Rather it starts from the concrete here and now, from given and actual human purposes, human needs, human responsibilities, and human relationships—and these, in turn, determine ethically what we need science for and how we ought to use it.

In addition to all this, it should be observed that there are theoretical as well as practical obstacles to developing a social program based on long-range predictions of the future development of science and technology, especially when it is predicated on their current status. As we all know, such predictions are notoriously unreliable, and risky. The only thing that we can be certain of is that there *will* be revolutionary new developments of some kind or other and that future science and technology will create as many new and unforeseen problems as they will solve old ones. As Herbert Simon points out, social planning with fixed goals is myopic. "[N]ew prospects are opened up at each step in the process of design. . . . We might be more optimistic if we recognized that we do not have to solve all of these (old) problems. Our essential task—a big enough one to be sure—is simply to keep open the options for the future, . . ."[5]

Mythologies and the Need for Discrimination

In addition to the rhetoric, there is a good deal of mythology in both ethicism and scientism. They take for granted, for example, that genes are the primary determinant of a person's future life, behavior, and happiness. Their attitude toward genetics and ethics is based on some form of what Richard Lewontin calls biological determinism. However, they are obviously unable to show that if someone has, say, Mozart's genes, he will automatically be able to write another *Marriage of Figaro*.There is precious little evidence for the claim that genetic makeup will replace culture as the determinant of future society.

Coupled with biological determinism, or perhaps at the root of it, is the ethical focus on genes instead of their product, mankind. The same sort of lopsided orientation toward mere biological existence may be observed in the mythological doctrine that endows subhuman life—for example, a blastocyst—with the attributes of personhood (in the ethical sense) and the rights of human persons.

I will refrain from further elaboration of these points. My own view of the nature of philosophy leads me to suppose that the most useful function of a philosopher in areas like genetical ethics is to try to untangle and clarify the underlying issues and conceptual problems. For this purpose, it is best to begin with a philosophical diagnosis of the questions that we want to ask about genetic engineering. In other words, we first need to ask questions about the questions themselves. For example, we need to ask about what might be called their etiology—that is, how they arise, why they are important, and what kinds of considerations (arguments) ought to be taken into account in answering them.

Questions: Problems versus Issues

Our first diagnostic task is to differentiate between problems and issues because, as I shall show presently, much confusion in genetical ethics is due to the failure to distinguish between these two kinds of questions. Although there is, perhaps, no standard usage with regard to the terms *problem* and *issue,* there is a standard meaning for each of these terms that reflects the logical distinctions on which I want to focus.

Problems in the sense that interests us here—namely, practical and ethical problems—are perplexing questions or situations that call for a practical solution. They take the form: What can we do about X? Problems exist where there is something wrong; they indicate evils to be overcome.[6] Thus, inflation is a problem, cancer is a problem, as well as terrorism, alcoholism, and teenage pregnancies. Genetic diseases like Tay-Sachs, Huntington's, or Down's syndrome are also problems in this sense. It is significant that the names of the problems are themselves the names of evils, as the examples just given show. So regarded, the satisfactory solution of a problem provides us with the know-how to cope with the evil—for example, how to eliminate or to mitigate it.

Thus, behind the perception of a situation as a problem are specific value assumptions and presuppositions about what is undesirable and evil. It also follows that if there is no evil, then there is no problem, and the question becomes a nonproblem. Needless to say, therefore, there is a kind of built-in relativity of problems: What is a problem for one person need not be a problem for someone else, either because it is not an evil suffered by him or because it is not viewed as an evil by him. A social problem involves an evil that affects society or that is viewed as such by society.

Issues, in contrast to problems, represent divisions of opinion, ethical disagreements, clashes of ideas and beliefs. They are the substance of controversies. Hence, an issue exists only when two (or more) parties disagree with each other on, say, a question of values. Perplexity is not essential to an issue.

When there is no ethical disagreement, as distinguished from an ethical perplexity, there is no ethical issue. Some issues that used to be issues are no longer issues. I assume, for example, that transubstantiation, over which bloody wars were fought a few centuries ago, is no longer a live issue. Dead issues are issues that people do not care about any longer. By the same token, in a homogeneous society where everyone shares the same ethical norms, there are no ethical issues, although there may, of course, be ethical problems—for example, questions about what to do.

In our own heterogeneous society, however, there are many controversial ethical issues. For example, in some regions divorce is an ethical issue, a controversial matter on which different parties disagree. Abortion is another example of an issue. Some people believe it is wrong, and others do not believe it is wrong. In vitro fertilization is also an ethical issue. On the one hand, for those who deny that anything is wrong, either with abortion or in vitro fertilization, these are nonproblems, ethically speaking. On the other hand, unless they are completely insensible, those who disapprove of abortion or in vitro fertilization are still faced with the original problems that these measures are used to solve—for example, teenage pregnancies and infertility. In that sense, abortion raises problems for an antiabortionist, although not for others. The other problems still exist regardless of the position that is taken on the ethical issues.

To sum up, problems are perplexities; there are perplexing ethical problems. But issues are controversial; there are controversial ethical issues. Clearly, what is perplexing need not be controversial, and what is controversial need not be perplexing. The logical structure of problems and issues is entirely different, not only as far as their preconditions are concerned but also as far as how one goes about resolving them is concerned. An issue is resolved by reaching agreement, either when one party convinces the other party or when they both give up their contrary opinions and adopt a third position on which they can both agree. Problems, as I have already indicated, are resolved when the means for eliminating or reducing the evil are found. It should be obvious from these considerations that the methods and logic of argumentation that are appropriate for dealing with controversies (issues) are essentially different from those that are appropriate for dealing with perplexities (problems).

Now it seems clear that the distinction between problems and issues is often completely overlooked in discussions of genetical ethics. Its relevance can be seen if we consider the following example. The use of a certain procedure for handling a genetic problem, say, amniocentesis plus abortion is, on the one hand, controversial—some think it wrong, and others think it not wrong. That is an issue. For one of the parties it is a nonproblem. For the other party, it is a problem only in an indirect sense. (Medical risks, of course, present a quite different sort of problem.)

On the other hand, a Down's syndrome infant (or fetus) presents a problem—what to do about it—for the occurrence of the affliction is an evil that we have to cope with, either by eliminating it or by adjusting to it in some way. Of course, the problems associated with Down's syndrome are more complicated than that because, at least at present, many perplexing problems relate to the care of persons afflicted with Down's syndrome. In general, however, the alleviation of hardship and suffering is the principal focus of the problems that concern us here.

Macro- and Microethics

At this point, we find it useful to distinguish between the problems of macroethics and those of microethics. Macroethics is concerned with the problems of society at large, what might be called social problems, and microethics is concerned with the problems of individuals, including their families.

Eugenics, I assume, would involve questions of macroethics, whereas genetic counseling would involve questions of microethics. It may be worthwhile to mention some of the problems of macroethics in order to bring out their differences from the problems of microethics.

Macroethics is concerned with what is good for society—that is, for people in general, rather than for a specific, assignable person or persons. Decisions relating to the good of society usually require collective decisions—that is, decisions in which many people, including those not directly affected, join. Typically, such decisions are about long-range plans and projects affecting numbers of people over a considerable time. They are often incorporated into legislation or into political decisions about financial aid or taxation. More often, perhaps, in our own society such decisions are made by the market—that is, what people are willing to buy and to sell. I am not sure this is the best way to settle matters of such moment.

Macroethics includes problems concerning the allocation of scarce resources. As far as genetic engineering is concerned, such allocations are relevant in two regards: first, how much and how should resources be allocated to support genetic research, and second, how much and how should resources be allocated to support expensive therapeutic measures? With regard to the latter, as a bystander, I suspect that in vitro fertilization will be such an expensive undertaking that society will not be able to afford it without having to sacrifice other more pressing needs. It is a luxury for the affluent.

The problem of allocation of resources is subject, of course, to cost-benefit analyses; and surely, if the costs are too high, society will not be able to undertake a project. However, in addition to cost-benefit considerations, considerations of justice and equity must be weighed in deciding questions about allocating resources. Here perhaps the primary consideration ought to be need; our resources should be directed at meeting human needs rather than, say, at satisfying human whims or wishes.

One of the most urgent problems of macroethics, in my opinion, is the problem of control—power and responsibility. Our society has not been very successful in controlling the exploitation of others by those in power. A real problem we must face is determining how to avoid the exploitation of genetic knowledge and genetic technology for economic profit without regard to social benefit and cost. If we leave it to the drug companies to decide whether or not to exploit new genetic disoveries, will it be for the good of society and of the individual, or will it be for the good of the drug companies?

My honest feeling is that the chief problems of macroethics are social and political, if not economic, and that we need to find solutions to them before we can start thinking about the problems of genetical ethics. Both the nay-sayers and the yea-sayers are off the track, for they seem to think that our problems will be solved either if we restrain the geneticists or if we give them full sway.

It may not be inappropriate to mention the near tragedy in the nuclear power plant at Three Mile Island. There does not seem to be any reason to quarrel with the idea of splitting atoms for energy, in principle and in theory. The real problem arises when organizations, private and public, have a vested interest in exploiting the idea for immediate gain. When we talk of social control, how are we going to arrange it so that the control is not by those who have a vested interest in pursuing one line that may be harmful or dangerous in the long run to many people, as against another line that will bring indisputable social benefits?

I shall leave these difficult problems with the admonition that we ought not to ignore them simply because other problems raised by genetic engineering are more spectacular and easier to understand.

Microethics and the Problem of Genetic Information

Many interesting philosophical problems concern the relationship between macroethics and microethics. Utilitarianism is principally a theory of macroethics; it really has very little to say about microethics. My own view is that macroethics, that is, the ethics of societies, reduces, perhaps in a complicated way, to the ethics of individuals. For present purposes, it is unnecessary to explore the relationship between these two kinds of ethics in further detail.

However, there is a particular problem here as far as genetics is concerned, because most of the basic laws and concepts of genetics relate to populations. After all, frequencies and risks refer logically to sets—that is, to populations rather than to individuals. Part of the problem of genetic counseling is logical—namely, the difficulty, perhaps impossibility, of translating frequencies within sets to some kind of analogue for an individual. This is a problem that I am unable to explore here.

Microethical problems, as I have indicated, focus on the problems facing individuals and their families—that is, on what Bentham calls assignable individuals. What should be done in particular situations about, say, genetic disease? Genetic counseling is, of course, the arena in which people confront the situations that are perplexing for them as individuals, quite apart from their wider social ramifications. For reasons I cannot explain here, I think that it is a mistake to mix social considerations with considerations that ought to be used to determine the decisions of individuals in particular cases. For example, the increase of numbers of persons afflicted with Down's syndrome may be a matter of general social concern, a problem for macroethics, but that consideration should not be used to determine what to do or how to treat a particular individual afflicted with Down's syndrome or to affect the decisions of a particular family concerning a fetus that has been determined to have a defective chromosome. What may be a good reason for an individual may not be a good reason for society, and vice versa. By the same token, the moral problems faced by particular individuals are bound to be quite different from those facing society at large.

Exactly how one should go about solving problems of microethics like those just mentioned is a long story. It is impossible for me to say anything substantial about it here. However, I want to warn against looking for a slot-machine ethics, one that will turn out answers to our ethical problems as a computer turns out answers to mathematical problems. Behind this kind of ethics is what John Dewey calls the quest for certainty, a will-o'-the-wisp as far as ethics is concerned. Furthermore, it represents a form of rule fetishism, the assumption that all we have to do is to find a rule for making ethical decisions of a certain type and then we can go about applying it. That is, of course, a cop-out, but worse than that, rule fetishism is one of the principal sources of inhumanity.

Rather than orienting our ethics in an absolutistic way toward rules, we should orient it toward what I call principles—that is, to what lies behind the rules and gives them life and meaning. Principles express features of those relationships between people that form the basis of morality.[7] As far as decision making is concerned, principles provide the touchstone for differentiating between morally worthy considerations and those that are not. Unlike rules, however, they are flexible and adaptive. With the rapidly changing conditions of life in the modern world, we can no longer afford to base our ethics simply on rules.[8]

The principal consideration in microethics—that is, the primary moral principle—relates to the autonomy of the individual or, as I prefer to put it, his moral integrity. His moral integrity also extends to his relations with others. Moral integrity requires having knowledge and information about oneself and one's relations with others. So, I should argue, the imparting of genetic information to those concerned is not only necessary for making the best possible decisions but also is a necessary element in moral relationships between persons. The deliberate withholding of information about a person from him constitutes an affront to his dignity as a person, his moral integrity. Furthermore, it undermines the integrity of his relationships with others, say, his family and his doctor. It destroys both integrity and trust.[9]

Genetical ethics at the microlevel essentially comprises many different kinds of problems and issues. In particular, areas like genetic counseling and genetic screening give rise to numerous ethical questions—namely, questions relating to the handling of detailed genetic information about individuals: How is it to be obtained? How is it to be used? Who should have access to it? These questions are frequently formulated in terms of rights: Who has the right to this kind of genetic information? Who has the right to make decisions based on it?

The ethical problems and issues are not, however, as cut and dried as the reference to rights might suggest because the use of the concept of rights itself usually amounts to nothing more than indirectly bringing in black-and-white rules of the sort discussed earlier. Insofar as rights have any use or validity in this context, they are (or ought to be) derived from what I have called principles, in particular, the notion of moral integrity.[10] Resort to principles like these will lead to a deeper understanding of the questions of genetical microethics although, at the same time, it will make our task of dealing satisfactorily with them more complicated and difficult; but, then, why should one assume that the task of ethics is simpler and easier than that of, say, the sciences and other disciplines concerned with genetic problems?

Notes

1. Probably the most intellectually respectable collection of articles on the subject is to be found in the *Encyclopedia of Bioethics*. These articles

make it clear that the longer one reflects on ethical issues in genetics, the more complicated they become. See Warren Reich, ed., *Encyclopedia of Bioethics,* vol. 2 (New York: Free Press, 1978), pp. 513–578.

2. There is no need to identify individuals in each party. A comprehensive list may be found in Ted Howard and Jeremy Rifkin, *Who Should Play God?* (New York: Dell, 1977).

3. Suppositions are not part of ethics, I would argue, for as Aristotle said, we "deliberate about things that are in our power and can be done," and we deliberate only about what can be done "by our own efforts." (Aristotle, *Nichomachean Ethics,* 1112a30) That excludes most of the scenarios envisioned by these ideologues.

4. Jacques Monod, *Chance and Necessity* (New York: Random House, 1971), p. 180.

5. See Herbert A. Simon, *The Sciences of the Artificial,* 2nd ed. (Cambridge, Mass.: MIT Press, 1981), pp. 188, 191.

6. By *evil* I do not intend to suggest that they are moral evils. An evil is an undesirable state of affairs, something harmful or troublesome, a disaster or a calamity.

7. They are, in Kant's terms, embodiments of the categorical imperative.

8. I have set forth some of these ideas in an article entitled "Legalism and Medical Ethics," in *Contemporary Issues in Biomedical Ethics,* edited by John W. David; Barry Hoffmaster; Sarah Shorten, pp. 1–35 (Clifton, N.J.: Humana Press, 1979). A shortened version of this article appeared in the *Journal of Medicine and Philosophy* 4 (1979).

9. For further details on this point, see "The Ethics of Participation," in *Participation in Politics,* edited by J. Roland Pennock and John Chapman (New York: Atherton-Lieber, 1975).

10. I have explained this way of deriving rights in my discussion of the right to die in *Ethical Issues Relating to Life and Death* (New York: Oxford University Press, 1979), especially pp. 137–140.

9 Genetic Engineering: Sizing Up the Arguments in the Post–Louise Brown Era

Howard Brody

In 1979, when the topic of genetic engineering was first discussed at a Science and Morality conference at Nazareth College, the discussion was dominated by the recent birth in July 1978 of Louise Brown, the first baby born as a result of the technique of in vitro fertilization. Since that time there have been further advances in genetic technology. Several other children have been born as a result of in vitro fertilization, and public debate has accompanied the establishment of new centers where this technique will be applied in the future. Further advances have been reported in sperm-selection techniques to increase the chance of having a child of the desired sex. Recombinant DNA technologies are booming, with new reports of breakthroughs in the pharmaceutical industry to produce much needed biological chemicals by this technique. There have also been reports of preliminary trials of recombinant DNA technologies for the treatment of genetic disease in humans.

It is not clear that our philosophical understanding of these issues has kept pace with the technological advances. We have good reason to agree with John Ladd, who stated in chapter 8 that "the literature in genetical ethics is . . . a miscellany of confused and confusing themes, pompous utterances, mythologies, dreams, and fairy tales." The fault does not lie with ethics as a discipline. During the same period, we have seen significant advances in our understanding of other biomedical issues as philosophers have approached them with rigorous tools of ethical analysis; improved understanding of the relationship between active and passive euthanasia is only one example.[1] Professor Ladd made an admirable beginning at bringing some order to the genetic chaos. I hope in this presentation to further the work of constructing a sort of ethical road map to the field of genetic engineering.

The task at hand may seem doomed to failure from the start. Very different practices have been included by one authority or another under the

---

The material in this chapter is adapted from "Genetic Engineering," in *Ethical Decisions in Medicine,* 2d ed., by Howard Brody, chapter 15 (Boston: Little, Brown & Company, 1981).

general category of genetic engineering, in addition to the in vitro fertilization and recombinant-DNA technologies just mentioned. These include various approaches to selective mating, artificial insemination by donor, and cloning and parthenogenesis (so far documented only in lower animals). Added to these technologies already used or experimented with in the past are the future possibilities of the artificial placenta and more direct selection and manipulation of eggs and sperm for a specific genetic trait. Some of these technologies fall under the category of negative eugenics—that is, altering the genotype of the offspring in order to avoid genetic disease or deformity. Other intended uses of these technologies fall under positive eugenics—that is, the selection of the genotype of the offspring in order to produce a superior stock. Other technologies, at least as they have been used up to date, are eugenically neutral. For example, we assume that Louise Brown was genetically neither superior nor inferior to the child that her parents might have conceived had Mrs. Brown not suffered from blocked fallopian tubes. It might seem that no general ethical statements or conclusions could possibly apply across the entire range of technologies just mentioned. Only case-by-case analysis of individual technologies could possibly produce useful insights. Indeed, the editors of the *Encyclopedia of Bioethics,* which Professor Ladd praised as having one of the few useful analyses of this subject area, were unable to produce a single general article covering this entire range of topics.[2] I believe, however, that despite the many differences in the natures of the technologies, and the uses to which they may be put, certain general comments can be useful, since the same sets of arguments tend to recur in debates about different technologies. The only way to prove this point, of course, is to proceed with the discussion.

In what follows, I will be working from the presupposition that ethics is useful even when it does not lead unerringly to a single statement of indisputable truth. There are many areas of biomedical ethics in which several alternative positions, each rationally supported by sound arguments, compete for our allegiance without our being able to give irrefutable grounds to choose one over the other. I believe that ethical tools of analysis serve us well in such cases even if they merely allow us to clear the field of the obviously erroneous arguments and to eliminate the weakest of the contenders before the major conflict is joined—that is, even when ethics cannot lead directly to truth, it can still help us by detecting error.

We can now turn to a reasonably comprehensive review of the different arguments that have been raised with regard to genetic engineering technologies. These arguments tend to sort themselves into two categories, depending upon whether they most commonly appear in defense of or against the implementation of the different technologies. Three arguments typically appear on the pro side:

1. *Utilitarian argument:* Genetic engineering entails some risks, but potential benefits to mankind in the long run outweigh these risks.
2. *Rationality argument:* Since our rationality is what makes us truly human, new technologies, allowing us to replace chance selection with rational choice, are to be welcomed.
3. *Genetic-decay argument:* We now engage in negative genetic engineering by keeping alive people with genetic diseases and allowing them to reproduce. This pollution of our gene pool requires an active policy to rectify.

Another set of arguments typically appears on the con side:

1. *Unnatural argument:* Genetic engineering replaces natural procreation, within a context of human relationships, with a sterile manufacturing process in the laboratory.
2. *Argument from uncertainty:* Since we cannot be certain in advance that genetic engineering will not lead to disastrous consequences, we should never begin to apply such technologies.
3. *Argument from ignorance:* Since we remain largely ignorant about possible long-range consequences of genetic manipulation, we should proceed very slowly before we start meddling with complex biological systems.
4. *Bad-values argument:* The desire to change the genetic makeup of our offspring embodies social values that have many undesirable features and consequences. Technology to implement such desires will encourage these bad values to the detriment of more positive social values.
5. *Future-abuse argument:* Genetic engineering techniques may be developed for totally good motives but may still be misused by future Nazilike tyrants to serve as instruments of repression and exploitation.
6. *Resource-allocation argument:* While genetic engineering technologies may serve worthwhile social goals with reasonable risks, they may be tremendously expensive to develop. Such resources may better be applied to higher priority health needs where low-cost technology already exists.
7. *Unethical-means argument:* Even if the ends of genetic engineering are ethically sound, the means to develop these technologies are immoral—including killing fertilized eggs and other products of in vitro procedures and carrying out experimentation on children-to-be who cannot consent to it.

This list is certainly not exhaustive. I have eliminated some extremely simplistic arguments such as the view that genetic engineering is to be totally banned because it amounts to playing God or the view that it is to be supported in every possible way because everything that is technologically possible ought to be implemented automatically. I would hope that we have reached the stage in our general understanding in the field of bioethics at which such arguments would no longer appeal to any of us. I have also omitted arguments that make reference to values or beliefs that are restricted to one particular religious group. Many of the arguments I have listed have been put forth most forcefully by theological ethicists, but they depend for their force on rational arguments that transcend the dictates of any one particular religion.

If all the arguments in the preceding list are equally weighty, we clearly have a difficult task in evaluating any proposed genetic technology. Conversely, if we can eliminate some of the weaker arguments, our job at arriving at sound ethical positions might be considerably simplified. We might begin by eliminating two of the weakest arguments from the list. The argument from uncertainty, by requiring absolute knowledge before we would be allowed to act, seems to place totally unrealistic and impossible conditions upon us, given the necessarily imperfect nature of medical and scientific knowledge. Taken to its extreme, this argument would reduce to the rule: Never do anything for the first time. The argument thus seems to put itself squarely against any legitimate change or progress. We may also find ourselves having little sympathy for the unnatural argument, since some form of this argument has been used historically to oppose many desirable changes, including the use of anesthesia to allay the pain of childbirth. This argument usually embodies hidden and dubious moral assumptions about which behaviors are natural and which are not and thus cannot be rationally evaluated until those hidden assumptions are made clear.

Another argument that cannot bear much scrutiny is the rationality argument, which has been put forth most forcefully by Joseph Fletcher.[3] The problem with this argument is that rationality appears to be just one of many important human characteristics and not the single dominant characteristic that makes us truly human. (Elsewhere in his ethical writings, Fletcher makes considerable use of concepts such as Christian love that do not seem to coincide with rationality in any perfect way.) If the rationality argument is taken simplistically, it may easily lead one into the trap of the technological imperative of doing things simply because we have the technology to do so. It also ignores the fact that it is sometimes rational to allow chance and randomness to operate in preference to a system of deliberate control or manipulation. This may be the case when we have very imperfect knowledge of the consequences of our actions or when considerations of justice militate against the use of deliberate controls. An economist, for example, who suggests that the free-market economy is superior to some system of wage and price controls may be proposing something that is either sound or unsound from an economic standpoint, but he is not being irrational merely because he is willing to allow a certain amount of chance in his system.

Another argument that turns out, on scrutiny, to have much less force than might at first appear is the genetic-decay argument. Here the full evaluation of the argument depends much less on ethical reasoning than it does on scientific fact. Geneticists such as Crow have demonstrated the fallacy of assuming that most cases of successful treatment of genetically determined disease have any major impact on the human gene pool.[4] Consider, for example, a single recessive gene that currently has a frequency in the population of 0.5 percent and that currently kills the individual who has the homozygous condition for that gene before he or she is able to reach reproductive age. It may be calculated that if a new medical treatment is developed to allow the individual to reach reproductive age and experience normal fertility, it will still take seventy generations for the frequency of this gene to double so as to reach a 1 percent population frequency. For certain other conditions with different modes of inheritance other than the Mendelian recessive, the doubling time could conceivably be much shorter. Nevertheless, when one considers other factors such as the incidence of environmentally induced mutations, it can be seen that the influence on the human gene pool because of the successful medical treatment of disease is a very minor factor. It is true that in individual circumstances our past success in treating genetically determined disease has led to new medical problems. An interesting example is the woman who has reached adulthood with normal intelligence, despite having the homozygous condition for phenylketonuria, because she was successfully treated with a low-phenylalanine diet in early childhood. This mother now has a very high level of phenylalanine in her circulation. Her mature brain tissues are not affected by this, but if she should become pregnant, this will pose a serious risk of retardation for her unborn child.[5] However, it is important to notice that this in no way influences the human gene pool. If the child does, in fact, become retarded because of his mother's biochemical status, this retardation cannot be passed on to future generations. The geneticist Edmond Murphy, (see chapter 4) was probably thinking of the genetic-decay argument when he said, "By all means let us reflect soberly on the consequences of what we are doing in genetic counseling and the prevention of genetic disease, but let us get it into our heads once and for all that we are not talking about an emergency that demands immediate, heroic, and ruthless action."

Another argument that fades away under close scrutiny is the future-abuse argument. As an example of how this argument falls down, we might look at the debate over cloning, since this technology seems to have captured the special fancy of popular science writers and newspaper columnists and to have given rise to some especially silly speculations. Ironically, the people who express dismay over the possibilities attendant upon cloning usually have equally inadequate views of the past and of the future. They are guilty of bad history because they seem to feel that it was precisely by genetic manipulation that past tyrants such as Hitler were able to do their dirty work. They forget the fact that the selective mating attempted by the Nazis to breed their superior race was only a very small part of the horrors of that political movement and that Hitler was able to achieve much more by manipulating the environment through his propaganda techniques and mass-extermination programs. If Hitler had had any cost-benefit analysts working for him, they would have told him quickly to forget all about genetic manipulation and stick with environmental manipulation because it had a much greater payoff. The cloning enthusiasts are also guilty of a flawed view of the future when they attempt to predict exactly what will happen if, for example, one were to clone 100 copies of Einstein or of Hitler. First, they forget that the personality of the individual is the result of a complex interaction between genes and environment and that it is not dependent upon genes only. The chances are very good that these 100 clones in either case would be 100 very different people. Some of the Einsteins would probably do very poorly at math and physics, and some of the Hitlers might turn out to be quite nice folks. We might also forget that these 100 clones, even though they were genetically identical to some other person, would still be people in their own right and would presumably have the same rights that people have now. They would stand in the same moral relationship to the person from whose cell they were cloned, as would two identical twins to each other. For this reason, if I chose to clone myself, so as to have a double from whom I could later obtain organs for transplant in case I suffered kidney failure, I would be forgetting that I would have no right to take away an organ from that clone without that clone's consent. For that matter, if the clone suffered kidney failure first, he would have every bit as much right to demand one of my kidneys as I would to demand one of his. Of course, in that future society, we might come to such a pass that human rights will no longer be respected and that routine violations of other people's personhood and bodily integrity would be the norm. If that occurs, however, it will be the decay of social justice that makes that future society repugnant to us and not the fact that the cloning of humans is technologically possible. There is certainly no necessary causal connection between the technology of cloning and the decay in our respect for the rights of others.

In summary, apparently one need not fear very much from the possible abuses of cloning, just as one cannot hope for very much in the possible advantages of cloning human beings. The cloning example is one of the more extreme cases, and the future-abuse argument, when applied to other technologies, may be more persuasive. As a rule, however, when the future-abuse argument does raise valid ethical points, these considerations can be stated more clearly as part of the argument from ignorance or the bad-values argument, to which we will turn later.

An argument that requires considerably more detailed analysis is the unethical-means argument. This argument has served as the major deontological objection to genetic engineering—that is, objections that are based on a priori moral considerations and not on the consequences that arise from the technology. Sometimes the unethical-means argument is tied directly to specific religious beliefs. For example, artificial insemination might be condemned because, even though the end result, the birth of a child, is good in itself, a necessary means to collect the sperm sample for insemination involves masturbation, which is religiously objectionable. Such arguments, of course, will appeal only to those who adhere to the relevant religious belief.

Other instances of the unethical-means argument transcend any specific religious teachings. For examples, we might look at arguments that have been raised against the technique of in vitro fertilization. One such argument is aimed at the practice of inducing multiple ovulation to insure that a number of eggs are obtained, thus increasing the chances of fertilization in the petri dish. However, if multiple eggs are fertilized, only one will be implanted in the woman's uterus, and the others would be disposed of. If one believes that human life begins at the time of fertilization, destroying these other fertilized ova constitutes the willful destruction of human life. One would then be using human life as a means to the desired goal of pregnancy, thus violating the ethical injunction of regarding others as ends and not as means only. However, if one does not believe that human life and human rights begin with the moment of conception, then this argument will be unpersuasive. It also happens that the technology of in vitro fertilization has turned away from the practice of removing multiple eggs from the ovary for purely technical reasons (see chapter 6). Hence, this argument is of limited practical import at the present time.

Another form of the unethical-means argument, however, is not tied directly to arguments about the personhood or the rights of the fetus or embryo. This form of the argument has been put forth most forcefully by the Protestant ethicist Paul Ramsey.[6] Ramsey emphasizes the essentially experimental nature of in vitro fertilization techniques. Even if one has done extensive studies in animals, including nonhuman primates, one must always begin to apply these techniques to human beings for the first time. One can never completely rule out the possibility that some unanticipated harm resulting from the experimental techniques will occur for the child who will be born as a result of the technology. It is important to see that this harm pertains to the child-to-be, who clearly is a human being with rights, and not merely to the embryo or the fetus. Possibly one might detect harm caused to the embryo or fetus and prevent the birth of the child by an abortion, if this was morally acceptable. It would still be the case, however, that very subtle forms of the damage (for example, minor changes in intelligence) might not be detected and that the child might be allowed to be born with such a defect. According to Ramsey, this is necessarily an unethical activity since there is no way that the child-to-be could ever possibly consent to being a subject to such an experiment. Since it is the very nature of the activity surrounding in vitro fertilization to constitute unethical experimentation upon children-to-be, who are never capable of giving consent, the practice is morally flawed from its beginning.

This argument has considerable weight when applied to particular examples of the use of genetic technologies for reproduction. We would certainly condemn a couple or their physician who wanted to use a totally untried technology or one that had a very high chance of risk merely because of overwhelming desire to bear a child, without any regard for that child's future well-being. However, as the risks attendant upon the technology are known to decline, the argument becomes much less persuasive. This seems to be the case with in vitro fertilization since, even though the entire technology has been used in only a few live births, it is quite possible to estimate with high reliability the chance of damage occurring at each particular step involved in the technology (see chapter 6). It is not merely the case that this argument, like the argument from uncertainty, could be taken to ridiculous extremes so as to demand absolute knowledge before any human action could occur. It might still be objected that at least adults know that our knowledge is incomplete and go ahead with their eyes open, while unborn children could never possibly give consent to be acted on in this manner. Instead, the argument seems unpersuasive in part because it is in the nature of simply being a human child to have been born under risk-laden circumstances without any possibility for prior consent.

This case is misstated somewhat by Richard McCormick in his defense of Ramsey's position.[7] Consider a couple who know they both carry a recessive gene for a severe condition so that each child would have a 25 percent chance of being afflicted. If they choose to have children, one might argue, we have an example of the child-to-be running risks without any prior consent; and if this is all right, we cannot object to other children-to-be running different sorts of risks as a result of genetic technology. However, McCormick denies that this example would be all right. The ethical thing for this couple to do, he states, is not to conceive children; therefore, the 25 percent risk of their children-to-be could not be used as an argument to accept other risks for other children. However, it is clearly the case that Ramsey would extend his concern to much more minor risks. It is certainly the case that we do not insist that parents-to-be avoid every possible environmental factor—cigarettes, alcohol, possibly even one aspirin tablet—that has a distant possibility of producing some minor defect in the newborn child. In addition, we do not insist that parents, before they are allowed to conceive a child, show that they plan to rear that child in an optimal emotional and physical environment; and in fact, we allow people to marry and have children even though we know that their emotional ability to raise children properly is limited. In these cases, we feel very upset when tragedy strikes the child because of these risk factors, but we do not feel that a moral outrage was committed because of the impossibility of prior consent. The real likelihood of the damage occurring does make a difference in our ethical analysis.

In conclusion, I believe that Ramsey's use of the unethical-means argument is fundamentally flawed. It does, however, serve as a useful reminder that any genetic engineering policy that considers only the desires of the parents or the likely benefit to society and that ignores the impact of the technology on the child and on future generations is necessarily deficient from an ethical standpoint. Once again we will see that the later arguments we will be considering capture these considerations equally well.

It does not seem possible to refute in any conclusive way those arguments that now remain. We may now begin review of these other arguments to see how they can help us in determining the ethically appropriate uses of genetic engineering technology.

At first glance, the bad-values argument looks as if it is open to considerable abuse. It is certainly easy to go overboard with vague claims that any new technology will undermine social values or pose new threats to marriage and the family. These are clearly easy slogans for the conservative to hide behind when he is unable to think of any really legitimate reasons to oppose some social change. Nevertheless, values are important regulators of social interactions and must be taken into account in any assessment of the long-term consequences of our behavior, since changes in behavior can, in the long run, lead to changes in values. Furthermore, changes in social values are very subtle phenomena so that it is difficult to predict the value implications of our current behaviors in advance. This means that if we do take values seriously, we must be especially alert to very subtle changes and that, if we wait for clear-cut empirical evidence of a shift in values in the negative direction, it may already be too late to take corrective action.

So far, the literature on genetic engineering has been characterized by a number of forms of the bad-values argument. A fairly extreme statement comes from Marc Lappe and Peter Steinfels in connection with those technologies that would allow us to select the sex of our children: "A society which would choose male offspring in preference to females (or vice versa) is probably not one which is ready for the responsibility of assuming regulation of the balance between men and women."[8] This statement is somewhat extreme because it takes the form of a Catch 22: A society that wants to use a certain technology shows thereby that it is incapable of properly using the technology and could only show that it could use the technology responsibly by not wishing to use it. Nevertheless, there is a point worth looking at here. The eagerness to manipulate the numbers of children that are born of a certain sex suggests strong attitudes about the values of people of different sexes; and these are the same attitudes that might create considerable social problems and injustice when they become incorporated into social patterns of behavior. A number of other suggestions have been made about possible deleterious values that could arise from the implementation of genetic engineering. Ramsey, in the work quoted previously, makes much of the so-called manufacturing process that will replace the normal biological form of birth. This concern is questionable, however, even if it is not merely a resurrection of the unnatural argument in a new guise. For example, if it were taken to mean that a baby born as a result of artificia insemination with the husband's sperm is somehow less fully human th a baby born from normal sexual intercourse, then the argument itself cou lead to more deleterious social values than the technology it is aimed However, Ramsey also calls attention to the possibility of genetic engin ing's altering our social expectations with regard to normal and abnor individuals. In this, he seems on much firmer ground. It is much m persuasive to argue that, if we come to believe that genetic defects ar op to some quick technological fix, we might have much less patien for those members of our society who exhibit such defects and much less impulse to treat them in a humane manner and to allocate resources for their care. In a similar vein, some social critics have argued that we have already gone much too far down the road toward the therapeutic mentality of seeing all human and social problems as some form of sickness one can cure if one only consults the right expert or healer.[9] These trends are said to have had an unfortunate impact on our sense of personal responsibility and on our power to change ourselves and our future. I might think that if I am less knowledgeable or less kindly than I wish, I cannot be blamed or be expected to work to correct things; it may simply be that my parents failed to get the right sort of expert advice from geneticists when I was conceived. Finally, a whole cluster of possibly deleterious values surrounds the desire to select specific traits for one's children. It is hard to argue with the parents who desire to use a technology that seems to be very low risk and that has the promise of increasing the intelligence of their unborn children. It could be argued, after all, that this is not merely a matter of selfish pride, that the more intelligent one's child is, the more opportunities he or she will have for success in later life. However, when one turns this possibility to more specific traits such as physical appearance, athletic prowess, or sex, then the children apparently are being seen more and more as instruments for the satisfaction of the parents, rather than as ends in themselves. This possible shift in values would seem to have serious consequences for family relationships and for appropriate child-rearing practices in our society. As anyone attending a Little League baseball game can attest, it is already very difficult at times for parents to view their children as separate entities who have the right to decide their own goals in life, rather than merely as means for the parent's own ego gratification. As it becomes more possible to select in advance certain traits in one's children or, for that matter, to discard the various mistakes that might occur, one has to wonder whether parents will be even less inclined to treat their children as individuals and even more inclined to treat their children with harsh rejection if those children do not live up to the parent's preexisting expectations.

The bad-values argument, then, warns us to consider long-range value implications for society when we come to add up the risks and benefits of proposed genetic technologies. This can be done either in terms of negative values to be avoided or in terms of positive values to be supported. For example, Warren Reich speaking at Nazareth College in 1979, said, "Our behavior as regards in vitro fertilization will be right and good if it conforms to the model of the good parent, which includes acceptance of offspring and a bonding that embraces commitment, loyalty, and love" (see chapter 7).[10]

We may next consider the argument from ignorance, which appears at first glance to be almost identical to the argument from uncertainty. However, while the argument from uncertainty is an absolute prohibition against actions when we have less than perfect knowledge, the argument from ignorance merely states that we should be hesitant about making advances in technology in proportion to our uncertainty about the possible consequences. In particular, the argument from ignorance draws on our contemporary understanding of natural systems and the delicate way they are balanced, so that any intervention from the outside is more likely to upset the system than to benefit it. New knowledge in genetics supports this point of view as we learn how traits initially thought to be due to a single gene in a simplistic cause-and-effect manner are now known to be due to complex interactions among many genes involving the delicately timed turning on and turning off of specific genetic sequences in response to environmental stimuli. The further awareness that some of the changes we might conceivably introduce in population genetics may to some degree be irreversible is a further warning against rash action.

The argument from ignorance has played an important role in the debate over recent attempts at human gene therapy. This argument serves as an important counterbalance to the natural desire of geneticists to implement promising therapies quickly, given the very real suffering of the persons subject to these diseases and the great difficulty we have in treating these conditions with existing medical therapy. It might be argued that, so long as the research subjects are adults and give their innformed consent in full awareness of the unknowns that remain in this field, no ethical problem can arise. We must note, however, that informed consent becomes less meaningful as the scientists themselves are unable to gauge accurately the likely consequences. Anderson and Fletcher express these cautionary notes as follows:

> It is our view that knowledge in several areas should be obtained from studies in animals before gene transfer in human beings is ethically justified. Specifically, experiments in animals should demonstrate that the new gene can be put into the target cells and remain in them, that the new gene can be regulated appropriately, and that the presence of the new gene does not harm the cell. . . .
>
> [G]ene therapy in human beings offers . . . enormous good. . . . We hope this laudable goal will not be jeopardized by premature experiments.[11]

It is worth noting that Anderson and Fletcher base their objections on specific gaps in our information about gene therapy. Until more information is supplied to fill these gaps, it seems reasonably likely that risks of considerable consequence will remain unknown to us. Such might be the case, for instance, if the gene for normal hemoglobin introduced into a patient with sickle cell anemia via a virus entered not only the bone-marrow cells where the gene could affect hemoglobin synthesis but also other cells of the body where hemoglobin synthesis normally does not occur and where this foreign gene might interrupt the normal processes of gene regulation. This contrasts with criticisms that involve global rejection of all gene manipulation based on imagined risks that might or might not be of practical significance.

The next argument to be considered is the resource-allocation argument. It seems that this argument would have considerable force, especially when taken in conjunction with the bad-values argument and the argument from ignorance. As it is expressed, nothing is incoherent or ethically unsound about it. However, in recent years, its practical impact on the genetic engineering movement has been very limited. It turns out that most of the technologies that recently have been developed are not particularly expensive compared to more usual sorts of medical care. Furthermore, many of them, such as in vitro fertilization, have been developed, not because resources were allocated specifically for genetic engineering but as a byproduct of research directed at other problems such as human infertility and the avoidance of unwanted fetal deaths. If in the future it should be recommended that considerable shifts of research money should take place into the genetic engineering area, this argument might become increasingly important.

At this point in our discussion, we are left with four arguments that have withstood stringent ethical analysis: These are the utilitarian arguments, in favor of genetic engineering; and the argument from ignorance; the bad-values agrument; and the resource-allocation argument, on the opposite side. Since the utilitarian argument claims that we should proceed with the technology if in general the benefits outweigh the risks, and since the other three arguments go into great detail about possible risks that might outweigh potential benefits, these four are really all features of the same argument. We might wish to add one additional consideration, however. People concerned with the rights of individuals frequently find fault with utilitarian arguments if the argument appears to provide justification for increasing benefits of a larger number at the expense of a few, especially if those few have their rights violated in order for that greater benefit to take place. We saw a case of how this might occur earlier when we discussed the possibility that childless couples might be willing to undertake great risks in order to bear children, so long as these risks would be borne not by them, but by the unborn child. If we combine the four arguments and include a provision to address this concern, we might come up with the following statement: It is ethical to proceed with a proposed genetic engineering technology if in general the benefits outweigh the risks, provided that in assessing risks one is careful to include:

> Allowances for possible underestimation of risks based on gaps in current knowledge;
>
> Consideration of long-term value implications, not just short-range practical consequences;
>
> Alternative use of scarce resources;
>
> Provision that, as much as possible, risks are borne by those who will stand to gain the benefits.

This statement, of course, is in the form of a utilitarian or consequentialist ethical statement. As such it will be very unsatisfactory to those who are committed to a deontological form of ethical argument, who would perhaps claim that any ethical statement that makes no mention of a priori rights and duties cannot possibly do justice to this complex issue. However, of all the arguments we reviewed, the only one that expressed a clear objection to genetic engineering couched in deontological terms was the unethical-means argument. If we have been successful in showing the weaknesses of that argument, no other arguments in deontological form remain for us to consider. This would suggest that, as a general rule, genetic engineering technologies do not by their nature violate any basic rights or duties; and thus, even if we believe that rights or duties ought to be paramount in our ethical thinking, we are still free to judge them based on a combination of their benefits and risks. Of course, it is possible that any particular use of genetic engineering technologies will violate important rights and duties. Nothing that has been said here suggests that genetic engineering should be free from the ethical constraints that apply to other biomedical technologies. For example, we have had occasion several times to mention the experimental nature of these technologies, and hence the normal ethical guidelines regulating experimentation on human subjects would be applicable. Our argument merely suggests that, in a case in which a genetic engineering technology is used in an unethical manner, the circumstances of use and not the fact that the technology involves manipulation of genes will be the ethically problematical feature. We should also note that acceptance of the utilitarian approach to genetic engineering certainly does not force us to accept uncritically any new genetic engineering technology. At one point in discussing the ethics of in vitro fertilization and other new reproductive technologies, Richard McCormick seems to suggest that whether one will be for or against these applications depends upon whether one is a consequentialist or a deontologist ethicist.[12] However, this seems to misstate the case. Not only does our ethical position emphasize the need to take risks as well as benefits into account, but it also highlights specific areas of risk that might otherwise be missed—specifically, the long-range value consequences for society. Many of the new proposals for genetic engineering technologies or their immediate implementation would fail to meet the tests laid out in our ethical position, and acceptance of that ethical position would then have an effect of putting the brakes on a number of potential advances. To the extent that these brakes are based on the argument from ignorance, however, they may remain only temporary if new scientific knowledge shows the real risks to be less severe. If this approach is followed, the geneticist thwarted in his desire to make early use of new technologies will at least be stimulated to pursue his research so that, with more information based on animal studies and other sources, his technology can be approved for use in the future.

Of course, a full discussion of genetic engineering in all its complexities would require a list of specific ethical problems that attach to each of the individual technologies. We have not, for example, begun to deal with the question of what we ought to do when an unmarried couple present themselves with the request for in vitro fertilization, nor have we begun to scratch the surface of surrogate motherhood. If, however, we have brought some order out of the chaos of the many conflicting arguments that attach to genetic engineering in general, and if we have eliminated the least plausible arguments so that those that remain deserve to be taken seriously, we will have accomplished our objective for this presentation.

Notes

1. James Rachels, ''Active and Passive Euthanasia,'' *New England Journal of Medicine* 292 (1975):78–80; and Paul Menzel, ''Are Killing and Letting Die Morally Different in Medical Contexts?'' *Journal of Medicine and Philosophy* 4 (1979):269–93.

2. Roger L. Shinn, ''Gene Therapy: Ethical Issues,'' in *Encyclopedia of Bioethics,* 4 vols., edited by Warren T. Reich (New York: Free Press, 1978), pp. 521–527; and Richard McCormick, ''Reproductive Technologies: Ethical Issues,'' in ibid, pp. 1454–1463.

3. Joseph Fletcher, *The Ethics of Genetic Control* (Garden City, N.Y.: Anchor, 1974).

4. James F. Crow, ''Rates of Genetic Change under Selection,'' *Proceedings of the National Academy of Sciences* 59 (1961):655–661; and Crow, ''The Effects of a Changing Environment on Man's Genetic Future,'' *Bioscience* 21 (1971):107.

5. R.R. Lenke and H.L. Levy, ''Maternal Phenylketonuria and Hyperphenylalaninemia: An International Survey of Outcome of Untreated and Treated Pregnancies,'' *New England Journal of Medicine* 303 (1980):1202–1208.

6. Paul Ramsey, ''Shall We 'Reproduce'?'' *Journal of the American Medical Association,* 220 (1972):1346–1350, 1480–1485.

7. McCormick, ''Reproductive Technologies.''

8. Marc Lappe and Peter Steinfels, ''Choosing the Sex of Our Children,'' *Hastings Center Report* 4 (February 1974):1–4.

9. Christopher Lasch, *The Culture of Narcissism* (New York: W.W. Norton, 1979).

10. Warren T. Reich, ''Ethical Issues in In Vitro Fertilization and Fetal Experimentation,'' Conference on Science and Morality (Rochester, N.Y.: Nazareth College of Rochester, April 5, 1979). Reprinted with permission.

11. W.F. Anderson and J.C. Fletcher, ''Gene Therapy in Human Beings: When Is It Ethical to Begin?'' *New England Journal of Medicine* 303 (1980):1293–1297. Reprinted by permission of the *New England Journal of Medicine.*

12. McCormick, ''Reproductive Technologies.''

Part II Interventions during the Life Cycle: Chemical and Behavioral Interventions and Environmental Ethics

Chemical and Behavioral Interventions

The key to understanding the issues discussed in part II is a well-defined sense of the concepts employed. It is currently popular to talk about behavioral interventions as behavior control. We have deliberately chosen not to do so. There is considerable difference between behavior therapy and behavior control. Behavior therapy, unlike behavior control, which is described later, is one form of psychotherapy. It is a clinical practice based on relieving identified problems (usually limited in goal and quite specific) through professionally established techniques involving planned manipulation of an artificially constrained and constructed environment. Frequently, both the treatment goal and methods are agreed upon by patient and therapist before treatment begins.

There are two exceptions to the concept of a general patient population. In some patient populations, because of severe patient limitation in cognitive areas of intelligence or judgment (for example, in profound mental retardation), behavior-therapy techniques may be the only possibility for interaction. In another type of patient population, the patients enter into the therapy arrangement because of legal requirements or involuntary commitments. Even in this situation, however, some joint agreement during therapy is often possible. The emphasis is still on the patient's welfare, his choice to cooperate with proposed treatment, and his needs, which can modify treatment approach. While his choices are limited by the reality of the legal consequences of his previous behavior, such a person may not diverge too far from the real, if unspoken, restrictions on choice all of us have. The ideal of a free (autonomous) patient, citizen, or individual is probably a false description of the setting in which all of us make choices if, by free, we mean unrestricted, uncaused, uninfluenced. Similarly, the supposed total control within institutions may also be an inaccurate description of the realities of institutional life. In the same vein, the purpose of involuntary commitment is to return a patient to some minimum norm of functioning that will make therapy a joint endeavor, and part of that norm involves returning the ability to judge and choose. The purpose is not, as implied by behavior control, to program judgments and choices but to develop the capability to judge and choose. The fact that a criterion of rational, or nondestructive, choice is used is not coercion but reflects normal survival and socialization values shared by the human community. These interactions, whether agreed upon jointly or managed by the therapist, are physician-patient interactions, with all the medical and moral values of such a relationship.

Behavior control is a term used in the context of social control. It is associated with Brave New World future scenarios, Korean War prisoner brainwashing, Russian political activist thought reform, and Madison Avenue consumer conditioning. This is a far different population from the medical one. The purposes are associated with political control and thought censorship, with repressive social organization, or with economic manipulation. These are all social-level issues very susceptible to ideological or political interpretation. Individual good is not even a consideration, except as serving society's good. Those implementing such activities are not professionally in control of the goal or treatment; their own professional values are not incorporated into the overall plan since this is *not* a clinical situation but a social activity. The interaction is clearly not that of the physician-patient relationship but that of a social-control-system/member-of-society association.

How, then, has the literature come to equate behavior-therapy and social-control issues? We would like to suggest that such a confusion has resulted from an extremely broad definition of behavior change, allowing it to encompass education, parenting, therapy—in fact, most human exchanges. Under such a broad umbrella, physician behavior therapists and subliminal-techniques experimenters share the same issues. Recent attempts to make this definition more precise can only help separate the medical concerns from the social-learning context to the benefit of both.

The second reason for this lumping together of such disparate activities is something we would like to call the Clockwork Orange syndrome. This involves a faulty perception (1) of what behavior therapists do and can do and (2) what other forms of behavioral intervention that use chemical and electrode techniques really accomplish. Studies reported in the *Annual Review of Behavior Therapy* reveal that once an experimentally controlled environment is replaced by the real world, most outcomes begin to deteriorate. In fact, a therapy-effected change that holds up for a six-month period is considered highly successful. Most do not hold up even for that long in a natural environment. A number of difficult problems make the reality of behavior therapy quite different from that portrayed—namely, the difficulty of transferring behavior from an artificial and structured environment to a natural environment, the individual's personality as a continuing dynamic process subject to regression or change, the low yield of aversive conditioners or negative reinforcement, and the questionable applicability of conditioning to broad behavior patterns such as aggressiveness and general sexuality.

The more sensational techniques of electrode or chemical intervention have also been distorted. The classic example of electrode control is that of neurophysiologist José Delgado's stopping a raging bull by use of current applied through an implanted electrode. The interpretation of this has been that humans may be in danger of being similarly controlled. This conclusion persists even though other experts maintain that the most that can be said is that through electrode stimulation some activities can be deterred (that is, the bull was forced by the current to keep turning, although its other behavior indicated it would charge) and some states might be induced (the bull exhibited confusion and frustration). It appears that stimulating specific brain areas produces variable responses and affects too many different functions to be of use in behavior-control schemes. The brain is too interconnected a system to allow precise, specific, limited results from such stimulation.

The human brain is the most complex organ known. The average adult brain contains approximately 100 billion highly specialized cells called neurons organized into intricate structural networks. Within the array, an individual cell can be stimulated by up to hundreds or thousands of other neurons and can transfer a signal along to as many others. Transmission through the network of neurons is accomplished by electrical impulses that travel along individual cells and by chemical substances called neurotransmitters that can communicate across the gap (or synapse) between neurons. More than two dozen neurotransmitters are known (a small fraction of the number that probably exists), and these function to facilitate, moderate, or impede the transfer of the impulse through the system. Other brain chemicals may function as neurohormones that exhibit a more widespread effect or act as modifiers of metabolic activities that take place within neurons. The complex interplay of these structural and chemical elements is still poorly understood. Parkinson's disease and some forms of depression have already been associated with defects in the brain chemical signals, and many psychoactive drugs now available apparently function by repairing defects or imbalances in the neurochemical system. Even with these drugs, specific results cannot be guaranteed, nor can their effect, if any, be predicted with precision.

Finally, behavior is part of a larger system so that different environments can generate different behaviors from the same electrode or chemical stimulation. Nevertheless, behavior-control literature assumes that such technologies (electrodes, psychosurgery, psychoactive medication, controlled conditioning) will be able to manipulate, distort, and destroy people.

David Begelman, in chapter 10, mentions the restricted and extended senses of the expression *behavior modification*. He gives a historical overview, distinguishing the legal sense of behavior modification from the professional sense and pointing to the trend of increased legal involvement in these issues, replacing the hands-off legal tradition with a libertarian one. This has considerable impact on right-to-treatment/right-to-refuse-treatment issues. He then turns to a question that we indicated was one of the critical ones in developing medical ethics: Are behavioral concepts value free to begin with? Definitions of normality or abnormality are actually value choices or preferences, and Begelman cites examples of homosexuality and the whole question of the labeling of a defect or disease (for example, Down's syndrome) to illustrate this point. As he argues, ''value-judgmental and scientific concepts are so interwoven, it is fantasy to believe one can be skimmed from the other in order to examine each in some sort of pristine, unadulterated state.'' While Begelman defines behavior modification as an explanatory theory rather than a specific set of techniques, he does concentrate on the broader concept of behavior-control and social-level concerns.

Burr Eichelman gives an overview of the ethical issues involved in any psychiatric intervention in chapter 11, and he also gives a good feel for the healing context in which that intervention occurs. He gives alternative ethical schemes or tools for dealing with these issues—for example, respect for persons and justice—but cautions that ethical problems in clincial medicine may be the result of a ''tension between two ethical principles'' rather than between the ethical and the unethical. He discusses pharmacologic and behavioral intervention in terms of respect for persons, outlining three differing contexts: (1) active consent to treatment, (2) passive consent to treatment, and (3) objection to treatment (involuntary treatment). Eichelman argues that intervention to create or restore a state of self-protection is an ultimate act of respect—the attempt to restore human attributes. He also maintains that intervention is sometimes necessary to help the patient recognize the duties he would have as a rational or potentially rational human being (not to harm others or to infringe on the rights of others). For him, responsible treatment must encompass the full period of intervention and may require social or milieu interventions as well. The parallels to education he finds considerable (recall the broad definition of behavior control)—for example, the goal of education is to promote a self-protective, rational human being; it is enforced by society though not with a total environment; and its goal of creating self-protection is in tension with the concept of freedom. While Eichelman discusses social engineering issues (including Clockwork Orange), he wants to distinguish these as social, not medical, issues. Though intervention might still be generated out of a principle of respect, he wants to make clear that many usages of the term *behavior control* are not medical ethics issues and involve problems not directly relevant to those issues.

Andrew Sorensen, in chapter 12, addresses the issue of pharmacological intervention, using a methadone-maintenance program as an illustration for discussion. After a historical review of methadone- and 1-alpha-acetyl-methadol (LAAM) maintenance and detoxification programs, and an identification of the relationship of the cultural context to ethical issues in the treatment of drug addiction, he identifies those ethical issues. Sorensen argues that both the medicalization and criminalization of addiction affect the choice of addictive drug(s). For example, physicians are arbiters of the appropriateness of addiction to some drugs, and law-enforcement and judicial officers are arbiters of the appropriateness of addiction to many other drugs. Because of this determination, the ability to choose is restricted, and he argues that a conflict arises between citizens' freedom to choose and societal concerns about treatment and well-being. Furthermore, the addicts' right to refuse treatment conflicts with society's obligation to provide treatment, as pointed out in chapter 10. He concludes by recommending modification of our societal approach by affirming the right of people to be addicted, affirming the right not to be addicted, and weighting both rights. The increased medicalization and criminalization of addiction is not ethically neutral. The values reflected in these social trends should be noted and considered in light of the effects they have on the rights of citizens to choose addiction and, if addicted, treatment. This is a rights approach to ethics as well as a consideration of the ethics of medical intervention within a broader concept of behavior control.

Environmental Ethics

The use of psychoactive medications illustrates the difficulty of intervening in a complex system without obtaining side effects or results that were not intended. On a larger scale, the activities of members of an industrialized society within their environmental system produce similar difficulties. Even in simple societies, chemical substances posed problems. There are reasonable indications that wood-smoke contamination within an enclosed teepee could create health hazards. There are naturally occurring sources of toxic substances—for example, high levels of mercury in remote streams and in some fossilized fish. Most plants carry on chemical warfare with plant eaters, so apple seeds, pennyroyal, and sassafras can all carry significant risk, and even the peanut can be a source of deadly levels of aflatoxin. Still we have managed to multiply our environmental problems considerably. Over 4 million different chemicals are now known, with about 60,000 of these in wide use. In addition, we are accumulating waste products of industrial processes at an increased rate, and disposal of the toxic chemical waste products is a compounding issue.

The convenience of plastic carries a cost; the chemicals that help us live well carry some risk. We need to ask questions about cost-benefit as a tool for ethical analysis, about the whole notion of risk, and about strategies for making choices. Is balancing costs and benefits the way to make choices, and should this be done for individual concerns or with a utilitarian calculus? Are we inherently risk takers as part of our definition of ourselves? Is there a major ethical difference between voluntary risks and imposed or involuntary risks? Can science tell us what the risks are or determine how much risk we should assume, in terms of efficiency or trade-offs? Is our goal safety, well-being, or self-determination? Can we price values? Should we regulate risks through the political process, or should we let experts decide?

Leo Hetling approaches these questions in chapter 13 in terms of his professional concern: the safety of New York State drinking water. In the process, he describes some of the major difficulties with toxicological data and with deciding on strategies for collectively handling the problem. He puts risk into a context of common risks, and makes a distinction between voluntary and involuntary risks. Contaminants in our drinking water pose involuntary risks and, for that reason, are to be considered very seriously in terms of macroethics, or collective ethics. Hetling includes a table of estimates that shows that the assumption of even a very low risk (one in a million) in a large population involves choosing or allowing the death of some number of individuals and raises the question of the ethics involved in this risk assumption. His example of conversion to carbon-filtration facilities raises practical and ethical questions about trade-offs. The trade-offs, or competing values, issue is almost always involved in collective or social-level ethics: We want both pure and safe environments and a bountiful life-style.

Mark Sagoff, chapter 14, also begins by discussing risk, asking what we are really afraid of and why we routinely assume risks as individuals that we would almost physically resist accepting or being forced to accept on the social plane of decision making. His points illustrate the very real difference between individual-level and social-level ethics. The problem of whether it is actually possible to develop an ethical system to handle both levels, or whether freedom, autonomy, or justice concepts can even in theory resolve conflicts between these levels, is not actually addressed, but perhaps an optimistic solution is assumed in terms of the operation of the political process. Certainly, Sagoff rejects the cost-benefit modern economic version of utilitarianism, primarily on the grounds that choosing a free-market model to determine our human choices or values makes little sense and that using such a model where no market system actually does exist is an ethically meaningless exercise. Sagoff argues that the real ethical concerns are not efficiency, safety, fear, or even welfare (benefit) but resentment at unfair imposition of risks that are not self-determined. For him, the ultimate question is not how much people would be willing to pay as individuals (a descriptive and therefore scientific statement), but how much society ought to pay per life saved, as determined by juries of average citizens (which Sagoff says is not only calculated individual *shares,* but a logically distinct vote on the *policies* and *principles* involved.) He argues that we have reason to resent rather than fear these hazards. What we should fear, he argues (based on a right perspective), is the conversion of entitlements created by law to "wants" and of rights (based on common understandings of mortality) to "benefits."

10 Ethical Issues of Behavior Modification

David A. Begelman

Discussions of ethical issues of behavior modification are bound to be unclear in advance of a definition of the field, yet such a definition is not easy to come by. Those developed by practitioners in the field will differ substantially from definitions supplied by lawyers, journals of opinion, educationists, or the media. In law, behavior modification is the administration of any therapeutic technique that produces drastic behavior change in a relatively brief period of time.[1] Where the primary focus is on constitutional issues pertaining to the involuntary status of patients or prisoners, legal definitions may be heedless of those observed among behavioral scientists. For example, a behavioral psychologist may frown on the identification of behavior modification with biomedical interventions such as psychosurgery, electroconvulsive therapy (ECT), or chemotherapy. Likewise, a neurosurgeon or psychiatrist may react sharply to being classified as a behavior modifier. There is a point to such disavowals. The physician who feels he adopts a sophisticated approach to target neurosurgery objects to having his activities curtailed because he is lumped together with professionals reported to use cattle prods on retarded individuals. Similarly, the behavior therapist whose approach to treatment involves a humane application of operant principles rejects identification with aversive methods such as those bringing about irreversible neural damage or tardive dyskinesia. He believes these methods are the prerogatives of physicians who have defaulted on the opportunity to develop less drastic approaches to severe behavior disorder.

If the law overlooks distinctions observed by professionals, it nonetheless can illuminate other kinds of similarities among diverse techniques it classifies as behavior modification. In this respect, the law is a kind of leveler—it rides roughshod over categories kept separate as a matter of course in the professions. Accordingly, it is immaterial to a ruling like *Rennie* v. *Klein* whether the behavior modification involved was administered by this or that professional, of this or that theoretical persuasion, using this or that technique or approach.[2] What matters is the issue of the right to refuse treatment of whatever kind, however administered. It matters little to a ruling like *Kaimowitz* v. *Michigan Department of Mental Health* how established a psychiatric-modality psychosurgery happens to be; it is still legally experimental.[3] It is irrelevant to the spirit of a ruling like *Knecht* v. *Gilman* that vomiting-inducing drugs ingested contingent on unacceptable behavior are called treatment by mental health professionals.[4] They still come under Eighth Amendment prohibitions against cruel and unusual punishment. It is also irrelevant whether the proponents of direct analysis trace John Rosen's methods back to Freud; they may still constitute assault and battery if a court so rules, as indeed it did.

A definition of behavior modification is hard to come by for quite another reason. At present, there are two commonly observed senses of the expression that are exclusive of each other, as well as unclear. The first of these, the restricted sense, embraces efforts that while regulated by law, are techniques of a professional establishment. The second sense of the expression, the extended one, covers much more. In fact, it refers to a remarkably diverse category of procedures of interpersonal influence drawing societal attention. In this group may be listed the techniques of brainwashing, Hare Krishna, the Reverend Moon, L. Ron Hubbard, Werner Erhard, the Symbionese Liberation Army (SLA), and the Reverend Jim Jones, to name a few. Of course, these two categories of behavior modification cannot be neatly separated. Is the individual who trains his dog to sit up and beg practicing the restricted or extended sense of behavior modification if his efforts quite accidentally coincide with well-known principles?

Because they are effective in achieving their stated aim, practices may evolve that coincide with the calculated, intentional use of behavior modification principles by psychologists. A person can learn to train a child or a dog effectively without so much as cracking open a book or taking a college psychology course. A political leader or public-relations expert may engage in effective ways of manipulating opinion without being able to describe what he does in theoretical terms. A person who practices behavior modification may not be a behavior modifier, if by this we mean operating out of some conceptual scheme of psychology. Charles Manson was no behavioral engineer in the manner of Frazier of *Walden Two* when it came to manipulating the members of his clan.[5] He was hardly ineffective in this role, however, and it is possible to bring the conceptual scheme of behaviorism to fit what he achieved in order to explain it. Two colleagues of mine have recently done just this in attempting to explain the effect of Erhard's est training by concepts gleaned from operant psychology.[6]

If a conceptual scheme like those we find in theoretical psychology can fit the actions of sundry individuals who practice successful forms of interpersonal influence, perhaps the real difference between operant psychologists and many others is not, in effect, what principles operate in the behavior change effort but what conceptual scheme a behavior change agent brings to bear on the explanation of success with it. Consequently, a therapist whose technique involves positive reinforcement is not necessarily a behavior modifier in our restricted sense. He can be a therapist of any persuasion or orientation since positive reinforcement is not the name of a specialized technique. It is a concept unique to a given way of looking at behavior, to a conceptual scheme, or theoretical system. A behavior modifier in the restricted sense is simply one who describes his activities in accordance with the theoretical framework in question, not someone who necessarily behaves differently from practitioners of other persuasions.[7] Thus, reinforcing verbal operants is not a therapeutic technique in competition with, say, the dynamic exploration of inner motives and feelings. It may only be a different way of conceptualizing such exploration. These considerations should put to rest the idea that behavior modification in the strict sense consists of a neat category or armamentarium of therapeutic tools developed by behavior therapists exclusively.

More to the point of this book are the ethical implications of the extended definition of behavior modification. The psychological effects of behavior modification efforts identified as brainwashing, conditioning, or programming are more often than not patterns deviating from cherished values or traditions—that is, undesirable behavior. A young person is not regarded as programmed by Billy Graham, William F. Buckley, Jr., or either of the two political parties but by the Unification Church of Reverend Moon. One is not conditioned by parents to perform such unremarkable acts as joining the Chamber of Commerce, the voluntary fire department or the PTA; only to out-of-character achievements such as political assassination, leaving the church, rejecting one's family, speaking an alien language like Sanskrit, or resorting to violence. *The Manchurian Candidate* was a film that embodied this type of fantasy. Occasionally, a person is seen as conditioned away from his personality flaws, as in the film *Çlockwork Orange*. However, the fictional depiction of such transformations usually exact their price, like being unable to listen to the theme of a Beethoven symphony without vomiting. Last, one is brainwashed by the SLA, as in the case of Patty Hearst, not by Jesus Christ, as in the case of Charles Colson and Eldridge Cleaver, who are usually characterized as reawakened or reborn. Ominous psychological processes always seem to mark the behavior of persons who reverse traditional values. Dramatic shifts are always seen as inexplicable, only providing they go in an unacceptable direction, and behavior modification as popularly understood, may only be a contemporary metaphor for palliating the bitter pill of unorthodoxy. St. Paul's conversion on the road to Damascus, like St. Augustine's, was as abrupt as the about-face of any heretic who suddenly defiled the sacraments or practiced sorcery. In the former case, however, a sinner is said to freely embrace the faith; in the latter case, one may be possessed by devils. One is never possessed by angels, apparently because there is little need to posit esoteric processes or complicated mechanisms for morally sanctioned behavior. Free will seems restored and diabolical possession or behavior modification no longer an influence when actions have the unmistakable, if somewhat prosaic, stamp of conventional morality.

Despite the fact that behavior modification may encompass a wide variety of behavior change efforts, the law has by and large been concerned with programs administered to captive populations. The most significant trend in evidence over the past decade has been the abandonment of a hands-off policy by the courts. The hands-off judicial stance dates back to the mid-nineteenth century, and it influenced rulings declaring the incarcerated patient or prisoner to be a "slave of the state" *(Ruffin* v. *Commonwealth)*.[8] In *Stroud* v. *Swope,* a 1951 ruling, a circuit court reaffirmed the tradition by rejecting the idea that the function of the courts was to review complaints of convicted criminals.[9] By the mid-1960s, judicial abandonment of the hands-off stance was clearly in evidence. In the 1970s, the so-called right-to-treatment legal tradition was firmly established. Not only have courts in the present decade served notice on penal or treatment institutions that their programs can come under judicial review, but they have also affirmed that adequate therapeutic programs are a necessary legal component in justifying the involuntary status of committed patients. In *Donaldson* v. *O'Connor,* a 1974 precedent, a federal court of appeals awarded damages to a hospitalized mental patient detained fifteen years in a Florida institution without treatment because he had refused ECT and chemotherapy on religious grounds.[10] The same right to treatment was extended to juveniles the same year in *Morales* v. *Turman.*[11] The foundation for later right-to-treatment rulings appears to have been established in *Rouse* v. *Cameron,* a 1966 district court ruling.[12]

Abandonment of the hands-off policy of the courts during the mid-1960s coincided with the judicial recognition of the adversarial nature of the relationship between involuntarily committed patients or prisoners and the institutions that housed them. It is no exaggeration to say that since the 1960s, courts have moved in the direction of maximizing the autonomy of patients against those acts stripping them of prerogatives enjoyed by ordinary citizens. In general, the law has been implementing a judicial policy that tends to embrace a libertarian point of view. Understandably, the American Civil Liberties Union (ACLU) has been supportive of the trend, as is documented by its publication of a guidebook on the subject.[13] Progress in what these authors refer to as a mental health bar is so fast paced, the Commission on the Mentally Disabled of the American Bar Association (ABA) has for some time found it necessary to publish a bimonthly *Mental Disability Law Reporter* to keep attorneys abreast of current legislative and judicial developments.

Application or interpretation of legal precedents is such an involved and time-consuming process, little attention is given to other implications of that body of law presently under discussion. For example, what factors are responsible for the libertarian thrust of rulings over the past decade? Historically speaking, two very different kinds of processes are at work. One is a conceptual process, the other is an historical one. The former process is implicit within a body of law itself, although of less visibility at certain points in time than it is later on. I refer here to those implications of a body of law that in and of themselves make certain legal rulings practically inevitable unless they are reversed. For example, a right to treatment as a constitutional reality has, as we have seen, already been firmly established in law. However, the courts have also been laying the foundations of the constitutional right to refuse treatment. The most recent important affirmation of this judicial stance was Judge Stanley Brotman's ruling in *Rennie* v. *Klein* in November 1978. Judge Brotman enjoined physicians from forcing chemotherapy on an unwilling patient. We can, of course, speculate to what extent the drug in question, prolixin hydrochloride, constituted an element that was pushing the court in the libertarian direction. The chief advantage of this agent is that its tranquilizing effect can be released slowly by means of an intramuscular injection so patients otherwise unable to assume the responsibility of adhering to a daily regimen can receive the same benefits through a single administration. However, coercive therapy in this case raises the prospect of physicians' forcibly injecting unwilling victims, a somewhat repugnant image to a judiciary already partial to libertarian ideals. The right to refuse treatment is a legal posture lagging considerably behind right to treatment. A mere handful of precedents can be cited before *Rennie* v. *Klein*. In *Winters* v. *Miller,* for example, a federal district court ruling of 1969, plaintiff argued that coercive chemotherapy by psychiatrists at Bellevue Hospital violated her rights as a Christian Scientist under the First and Fourteenth Amendments.[14] Although the court rejected plaintiff's case, the decision was reversed on appeals two years later. In an earlier ruling, *Peek* v. *Ciccone,* a federal court had rejected plaintiff's claim that the coerced administration of a tranquilizing medication constituted a violation of the constitutional strictures against cruel and unusual punishment.[15] However, in *Rennie* v. *Klein,* the die has been cast in favor of the right to refuse treatment, and other suits have been initiated in major cities on the East and West Coasts, as well as in the South.

The conceptual process responsible for the direction the law takes might already be at work in the relationship among three judicial realities: the right to treatment, the right to refuse treatment, and the need for involuntary hospitalization or detention. All three are on a collision course, the upshot of which will be the rejection of one or more of these principles; if the hypothetical patient who refuses treatment does not, by virtue of this discretionary right, receive any treatment at all, the legal justification of an involuntary status under right-to-treatment laws has been undermined—that is, under the provisions of right-to-treatment precedents, institutions cannot deprive patients of their liberty *unless* a treatment program is available. However, a refusal to participate in such a program effectively nullifies this possibility. Furthermore, legal acknowledgment of the necessity of involuntary hospitalization, either through probation or physician's certificate, is by and large a need in relation to those unwilling to be so hospitalized. The abolitionist platform of libertarians like Thomas Szasz may thus ultimately be assured by the legal crunch in the relationship between right-to-treatment and right-to-refuse-treatment rulings. It may not even be necessary for libertarians to seek direct support for abolition of involuntary commitment if they are so inclined. This may already be the upshot of the tension among legal traditions in which it will eventually get sacrificed.

Conversely, developments in law bear the unmistakable stamp of events external to it. Just as right-to-treatment or right-to-refuse-treatment rulings acknowledge the sometime adversary nature of the relationship between patients and institutions, so too has a similar perception of the relationship between parents and their children crept into law. A number of rulings including *Bartley* v. *Kremens* recognize the right of children to resist commitment procedures initiated by their parents.[16] I have commented elsewhere that such a judicial stance is in turn predicated on the legal assumption that what is in the interest of the child can be ungummed from what is good for the family, while sociological trends such as the dissolution of family ties may pave the way for the possibility of these perceptions.[17] Likewise, it would be naive to assume that juridical sympathy with the idea of abortion as a maternal right, as in *Roe* v. *Wade*,[18] was uninfluenced by the impact of the women's movement or the abatement of church-related antiabortion campaigns in the wake of Catholic disaffection with the encyclical *Humanae Vitae* of Pope Paul VI.[19] Nor is it farfetched to suppose that defeat of the Equal Rights Amendment (ERA), together with the increased activism of antiabortion groups, will tend to influence, if not actually reverse, the direction of legal thinking on the matter of abortion, whatever its final disposition. Not only is ethical and legal thinking a sensitive barometer of the thrust of social sentiment, it may be also that we are all prisoners of history, to some extent. What we are prepared to view as an ethical or legal problem might depend to a great extent on the pull of social forces that give shape to our moral perceptions.[20] The outrageous professional practices of yesteryear may seem this way only because of the considerable advantage of historical hindsight. In the future, similar indictments of therapeutic efforts that seem innocuous to us now may serve to instill a greater measure of humility in us about our professional forebears.

At first, professional behavior change agents did not take lightly to legal review of their activities. A proprietary feeling about one's professional area and jurisdictional compass is perhaps inevitable, and the threat of moratoriums or curtailment weighed heavily on the shoulders of colleagues in the mental health profession. Despite their resentment, I cannot help but reflect on the poetic irony of it all: My colleagues are extraordinarily reactive when the courts seek to regulate their activities, whereas they are oddly silent when the process goes in the reverse direction. You will find very few psychologists and psychiatrists who blink an eyelash over what Kittrie calls the divestment of criminal law.[21] This is the tradition that, over the years, has eroded the concept of legal responsibility. Right or wrong, the tradition has so complete a grip on the law that most people have practically identified the insanity defense as a psychiatric, not legal, determination. In addition, significant shifts in the criteria of insanity, going from the wild beast test to the M'Naghten rule to Judge Bazelon's ruling in the Monte Durham case, have accommodated nicely to the vagaries of psychiatric fad. For example, the M'Naghten rule, as Gendin has observed, was never intended as propounding criteria of mental disorder.[22] However, divestment of the criminal law works to blind us to distinctions—like the one between mental disorder in psychiatry and nonresponsibility in law—that once had an honored place in jurisprudence. Currently, determination of psychiatric status in many cases is practically synonymous with a defeat of the claim of criminal responsibility. This is different from construing such a status as merely another bit of significant information courts may rely on in determining legal guilt.

The insanity defense is but one illustration of the effect of psychology on the law. Divestment is actually a more generalized phenomenon, characterized by a parade of what strikes me as explanatory fictions: the unconscious wish, the compulsion, the subliminally based act, the brainwashing, the programming, the deprogramming, and others in the retinue of newly devised gimmicks for the death of responsibility. When a black ghetto youngster filches a trinket from a five-and-dime store, authorities have intimations of juvenile court proceedings and a diagnosis of antisocial character; but when the white dowager is caught red-handed by store detectives in the same act, an elaborate social mechanism of extrication is set in motion, with a diagnosis of kleptomania or compulsive stealing in the offing. The diagnosis is not made on the basis of any psychological process confirmed to be operating in similar cases; it is made because it is rich in exculpatory possibilities, especially for the socially prominent. As such, it is hardly a scientific act but more like a moral or political move. The author of a compulsive act (or an unconsciously determined one) comes to be seen as one whose relation to it was less than intimate. It is as if the basis for ownership of the act were undermined and in its place substituted a paramechanical myth. The act becomes an accretion as unrelated to the agent as a fungus is to his intellect.

Yet another kind of issue surfaces in discussions of the ethics of behavior modification. In the diagnosis and treatment of behavioral problems, we focus on a wide range of labeled conditions or patterns demanding our attention. Often, we feel that, if something has been labeled in a particular way, this qualifies it as a condition, where the force of the term *condition* is tied to the undesirability of the pattern. Originally, of course, the reverse process was responsible for why a pattern received a label at all. We would have never been prepared to affix specific labels unless the patterns come to be known as conditions or syndromes were initially adjudged as undesirable. Thus, we would never be tempted to observe that Down's syndrome must be an undesirable condition because it is caused by an extra chromosome, unless the latter is associated with an organismic pattern or state evaluated as unwanted according to social, moral, or political criteria. Hence, the appeal to the existence of biogenic defect or impairment in order to designate treatable or undesirable conditions, far from settling the issue, only begs it. The hypothetical neurobiologist of the future could never show that the abilities of geniuses like Mozart or Gauss sprang from a defect of the central nervous system. The very notion of a defect is conceptually wedded to a normative evaluation of the condition it determines.[23] Unless variables responsible for the endowments of Mozart and Gauss were genetically coupled with pathological traits, it is logically impossible such endowments were the result of a defect or impairment.

It is possible to designate Down's syndrome a defect, or philosophical evil in Professor Ladd's sense (see chapter 8), but an evil according to whose criteria? As we have seen in this book, parents of children with Down's syndrome object to such a classification. Perhaps this is because they have developed a perception of what others refer to as their child's impairment that for them cannot be as easily disengaged from other of his personal characteristics. Consequently, for such parents, designation of Down's syndrome as a philosophical evil in Professor Ladd's sense may be tantamount to a negative judgment of their child as a person. The difference here is not between viewpoints reflecting different conceptions of the facts. On the contrary, it seems to be a difference introduced by contrasting value systems.

When we come to the realm of overt behavior, the value-judgmental element in all attempts to classify and modify behavior can make clinical activity a complex and self-deceptive affair. For example, it is difficult to specify the criteria for why we judge this or that pattern a disorder, a clinical condition, a pathology, a defect, and so on. I believe many of these judgments stem from essentially moral decisions embedded deeply in human traditions. They do not evolve from purely empirical considerations.

As in the case of other forms of therapy, behavior therapy for homosexuals was offered by practitioners who classified it as a problem behavior.[24] Today, we are not so sure about this characterization, and the

Nomenclature Committee of the American Psychiatric Association (APA) has long since acted to remove homosexuality from the profession's list of psychiatric disturbances.[25] If the concept of a disorder is somehow based upon the empirical features of a behavior pattern, how, in the case of homosexuality, is it employed in a relatively decisive way at an earlier point in time, whereas employed in an uncertain way or not at all at a later point in time? My belief is that official psychiatric doctrine on homosexuality is simply an expression of moral sentiment disguised as scientific pronouncement. As such, it is vulnerable to all the influences of time, place, and history one can expect from such a phenomenon.

The doctrine that the behavioral sciences are value free may not even be true of the concepts that are its lifeblood, not to mention its applications. When ethicists raise questions concerning the legitimacy of this or that procedure, even they are often in the grip of a certain picture of the relationship between valuational issues and behavioral science. The picture is one of a conceptual disengagement between the two. Thus, an ethical issue or dilemma is perceived as an occasion for an interface between two otherwise distinct and separable levels of discourse: the factual and the moral judgmental. My own belief is that value-judgmental and scientific concepts are so interwoven that it is fantasy to believe one can be skimmed from the other in order to examine each in some sort of pristine, unadulterated state.[26]

Often, as has occasionally been the case in this book, a particular question is presupposed by ethicists: What is the place of value in a world of fact? It is not a question the behavioral scientist usually feels obliged to answer, unless nudged reluctantly to do so by forces he considers historically bothersome, although inevitable. His concessions to ethical demands or considerations as he secretly sees it represent nothing more than the rituals of accommodation to groups who would seriously hamper his inquiry if antagonized. The behavioral scientist may indeed be right about the emptiness of ethical concerns if what he takes to be ethics is simply the iteration of creeds or dogmas on the part of interest groups as vested as his. However, I do not mean by the value-impregnated nature of behavioral science the mere clash of parochial viewpoints. I mean by it the impossibility of separating the wheat of purely empirical concerns from the chaff of moral rules, assumptions, or viewpoints. Behavior modification, however we define it, will always be an arena in which the point is drummed home repeatedly. Here wisdom dictates I should do something other than commend the subject to you, but I am going to do precisely that.

Notes

1. D.A. Begelman, ''Ethical and Legal Issues of Behavior Modification,'' in *Progress in Behavior Modification,* vol. 1, edited by M. Hersen; R. Eisler; and P. Miller, pp. 159–189 (New York: Academic Press, 1975).

2. D.N.J. 9 November 1978.

3. 42 U.S.L. Week 2063 (Mich. Cir. Ct., Wayne Cty., 10 July 1973).

4. 488 F.2d 1136 (8th Cir. 1973).

5. B.F. Skinner, *Walden Two* (New York: Macmillan, 1948).

6. D.M. Baer and S.B. Stolz, ''A Description of the Erhard Seminars Training (est) in the Terms of Behavior Analysis,'' *Behaviorism* 6 (1978):45–70.

7. D.A. Begelman, ''Ethical Issues for the Developmentally Disabled,'' in *Current Trends for the Developmentally Disabled,* edited by M. Berkler; G. Bible; S.M. Boles; D.E.D. Deitz; and A.C. Repp, pp. 41–66 (Baltimore: University Park Press, 1978).

8. 62 Va. (21 Gratt.) 790, 796 (1871).

9. 187 F.2d 850 (9th Cir. 1951).

10. 493 F.2d 507 (5th Cir. 1974).

11. 383 F. Supp. 53 (E.D. Texas 1974).

12. 373 F.2d 451 (D.C. 1966).

13. B.J. Ennis and R.D. Emory, *The Rights of Mental Patients* (New York: Avon Books, 1978).

14. 306 F. Supp. 1158 (E.D.N.Y. 1969); on reversal, Winters v. Miller, 446 F.2d 65 (2d Cir. 1971).

15. 288 F. Supp. 329 (W.D. No. 1968).

16. 402 F. Supp. 1039 (E.D. Pa. 1975).

17. Begelman, ''Ethical Issues for the Developmentally Disabled.''

18. 410 U.S. (1973).

19. A.M. Greeley; W.C. McCready; and K. McCourt, *Catholic Schools in a Declining Church* (New York: Sheed and Ward, 1976).

20. Begelman, ''Ethical Issues for the Developmentally Disabled.''

21. N.N. Kittrie, *The Right to Be Different: Deviance and Enforced Therapy* (Baltimore: Johns Hopkins University Press, 1971).

22. S. Gendin, ''Insanity and Criminal Responsibility,'' *American Philosophical Quarterly* 10 (1973):99–110.

23. D.A. Begelman, ''Misnaming, Metaphors, the Medical Model and Some Muddles,'' *Psychiatry* 34 (1971):38–58.

24. M.P. Feldman and M.J. MacCulloch, *Homosexual Behavior: Therapy and Assessment* (New York: Pergamon Press, 1971).

25. The decision by the APA Nomenclature Committee to remove homosexuality from the *Diagnostic and Statistical Manual-II* was based

upon somewhat paradoxical arguments. Dr. Robert Spitzer, a spokesman for the committee, has defended its decision as scientifically sound, although admittedly influenced by Gay Liberation political pressure. In addition, the decision is a qualified one. According to the committee, homosexuality is not a disorder per se but is such only when the pattern causes an individual undue distress. Such criteria pose troublesome questions, however. For example, why designate paranoid delusions as disorders per se if they fail to create distress in the individual harboring them? What is the definition of an emotional disorder to begin with? Is heterosexuality a disorder if it gives the individual emotional problems? Do cancer and tuberculosis cease to be medical disorders in the unlikely event that activist groups bring pressure to bear on categorizing them otherwise?

26. An illustration of the grip of this view is the conviction, shared by Senator Robert Dole's congressional committee, that an objective investigation of contemporary cults is possible. However, the legitimacy of any such investigation hinges on the proper delimitation of its subject matter. What is its subject matter?—that is, what empirical rule of inquiry enables us to define an area of investigation including the Unification Church, Scientology, and Hare Krishna but not Masonry, Mormonism, Hassidic Jewry, and Roman Catholicism? Is the contemporary use of the term *cult* already laden with negative moral connotations? Is it a rhetorical device used by majority religionists to oppress minority religionists? Are accusations of programming a cryptic expression of the sour-grapes theme? Perhaps such paramechanical concepts are meant to remind us that teenagers do not turn to alien belief systems unless something has been done to them. If so, we have a modern version of the ancient theme of diabolical possession.

11 Ethical Issues of Pharmacologic and Behavioral Interventions in Psychiatry: Aspects of Education and Respect for Persons

Burr Eichelman

This chapter is written from the vantage point of a practicing psychiatrist. In community psychiatry, the therapist encounters many difficult clinical situations requiring decisions that must be based on incomplete clinical information and that involve major ethical dilemmas. Most often the mental health practitioner begins with the ethical problems posed by a specific case and works his way through to a general ethical concept that he later attempts to apply to subsequent cases. As therapies have become more effective, there has been increased concern on the part of our society, and increased introspection on the part of therapists, regarding the ethical elements involved in psychiatric treatment.

Psychiatric treatment incorporates elements of behavior control. Here behavior control is construed broadly, utilizing London's definition, to be "the ability to get someone to do one's bidding."[1] With the success of antipsychotic medications in modifying schizophrenic symptoms,[2] of antidepressants and electroconvulsive therapies in reversing depressive illness,[3] of lithium in controlling many manic psychoses,[4] and of token economies in activating and resocializing certain chronic schizophrenic populations,[5] there has arisen an interest and a concern regarding the use of these effective tools for behavior modification. Indeed, we should all recognize that if no techniques were effective in modifying behavior, there would be little debate about the ethical use of such techniques.

One fact should be stated at the outset of this discussion—namely, that the techniques in the mental health therapist's armamentarium are and will undoubtedly continue to be limited in their power to control the broad spectrum of an individual's behaviors. These techniques can modify behavior but cannot control the total life experience of an individual. Clinical psychiatry is a bit more modest today than when Watson challenged society to:

> Give me a dozen healthy infants, well-formed, and my own specified

> world to bring them up in and I'll guarantee to take any one at random and train him to become any type of specialist I might select.[6]

The recognition that psychiatry's tools for behavior modification are modulators of behavior, rather than complete controllers and dictators of behavior, may allow for a prudent and reasoned exploration of their ethical use.

Ethical Principles in Pharmacologic and Behavioral Interventions

In the past two decades, myriads of ethical issues such as proxy consent, right to treatment, right to refuse treatment, or the psychiatrist as a double agent have evolved.[7] These issues should be examined using broad overarching ethical principles such as respect for persons or justice.[8] Such ethical principles have their own long and elaborate philosophical history. When understood by the clinician, they can become touchstones or navigational aids in the difficult decision-making processes of a mental health practice that incorporates elements of behavior modification.

A first observation and consequence of using ethical principles is the discovery that many ethical dilemmas in medicine represent a tension between two ethical principles rather than an ethical and an unethical option. For instance, because of respect for persons or the principle of personal liberty, a court may be unwilling to intervene to prevent the squandering of a manic patient's resources; however, this may conflict with the ethical principle of distributive justice by placing unbalanced burdens upon family or the society in the later provisioning of food and shelter for such an individual.

A second observation that must be made in the course of a chapter concerning bioethical issues is that ethical issues and decisions are not necessarily legal decisions and that legal positions are not necessarily always consistent with ethical principles. A clinician may well be confronted with a situation where a breach of confidentiality or a decision to treat an objecting psychotic patient could be supported as being in harmony with certain interpretations of the principle of respect for persons yet may be in clear violation of existing state laws. In such circumstances, the clinician, if he acts contrary to the law, must be prepared to accept the consequences of his actions.[9]

It is now useful to examine the ethical principle of respect for persons in some detail, with the goal of applying it to pharmacologic and behavioral interventions in psychiatry. The principle of respect for persons was initially set forth by Kant.[10] Though it can be elaborated upon and developed along different lines, it essentially states that human beings, on the basis of their actuality of, or potential for, being rational (that is, self-affirming, self-determining, and rule following) merit respect. Respect encompasses, at the least, taking into account and weighing heavily the feelings, aspirations, and potentials of the individual with whom one is dealing. Some might also maintain that respect includes attempting to maximize the opportunities for the individual to achieve his potential and his goals. Respect for a person precludes treating that individual as an object, solely as a means to an end. Harris has expressed the corollaries of respect for persons in the following way:

> (i) [E]ach and every person should be regarded as worthy of consideration, and should be so treated; (ii) . . . no person should be regarded by another as a mere possession, or used as a mere instrument, or treated as a mere obstacle, to another's satisfaction; and (iii) . . . persons are not and ought never to be treated in any undertaking as mere expendables.[11]

Beginning with respect for persons, we can now move to the application of this ethical principle to the psychiatrist's decision making in treatment issues for his or her patients or potential patients. These current ethical issues are encompassed in the right to accept or refuse treatment.[12] For the patient and the therapist, treatment issues fall into three categories: (1) active consent to treatment, (2) passive consent to treatment, and (3) objection to treatment, with the option for involuntary treatment.

Active Consent to Treatment

There is little conflict with the initiation of treatment alliance between a therapist and patient for the purposes of behavior modification when contracted voluntarily by the patient; particularly this is true when the patient seeks a goal of specific symptom relief. Examples of such a contract would include the treatment of phobias by desensitization techniques[13]; the treatment of anxiety by psychotherapy, relaxation techniques, or medication; and the treatment of a thought-disordered illness such as schizophrenia, voluntarily, with the purpose of obtaining clearer thinking or the removal of hallucinations. Such treatment contracts are formed with the attempt to maximize the opportunities for the patient to achieve his potential and his goals.

Passive Consent to Treatment

For the situations of passive consent and objection to treatment, it is necessary to examine more fully ways of proffering respect. Jonsen and Eichelman have recognized healthy persons as being not only:

> [a]utonomous and responsible but also . . . self-protecting. Persons must be able to devise plans, pursue them, modify them, engage the interest and win the support of others for them—all against the adversity of circumstances, the threats of environment, and the self-interest of others. We call the ability to do this the attribute of self-protection; autonomy and responsibility are contingent upon the ability to mobilize one's inner resources and outer circumstance to attain a goal which others will acknowledge.[14]

This is in contrast with severely disturbed mentally ill individuals:

> Psychotic patients suffer from serious deficits in self-protection; they are prey to bizarre ideation and rampant affectivity; they are oppressed by unrelenting doom. They are incapable of rallying inner strength, and by their strangeness they repel, without wanting to, the proffered support of familiars. They are radically vulnerable. It is the ultimate act of respect, the principle of psychiatric ethics, to attempt to restore or create self-protection.[15]

The process of affording maximal respect for an individual through the process of creating or restoring self-protection is a key element in the decision to treat the passively consenting patient. Examples of such individuals are wide ranging. There may be the grossly psychotic and disorganized individual who somehow manages to arrive by his own accord at a hospital emergency room but who then becomes mute and perhaps even catatonic, failing to provide a medical history or to assist in the development of his own treatment plan. Another is illustrated by the mother with a postpartum depression who sits in a chair at home, unresponsive to the needs of her newborn infant yet unable to initiate any intervention herself, who is brought to the hospital by her spouse while she remains totally passive in the treatment process. In both these situations, the patients are extremely vulnerable. As a consequence of their mental states, they are buffeted about internally and have surrendered their right to maintain their own self-affirming, self-protective state. Respect for them encourages intervention.

Objection to Treatment, or Involuntary Treatment

The most difficult area of intervention involves treatment and behavior modification of the objecting individual. In many circumstances, the refusal to accept treatment may well be an act of self-protection by the patient. However, this chapter argues that a person is no longer acting in a self-protective manner if, as a result of a mental illness, he exhibits behavior dangerous to self or others or is gravely disabled as evidenced by his inability to provide for food, clothing, or shelter. Intervention at this stage, even against the individual's expressed wish, becomes the optimal way of offering respect to that person. To allow a psychotically depressed individual the freedom to seek suicide on the basis of delusional beliefs negates and denies the potential within that person to a self-protective, functioning state. Rather, it could be argued that one is treating the person as an expendable commodity if one fails to intervene in suicidal behavior. One worries at times whether the civil-liberties lawyer who considers the victory of his court case above the needs of the patient to return to a nonpsychotic, self-protective state is using the patient as an instrument (an expendable object, tool, or pawn) in an attempt to maintain an extreme concept of freedom.

Further, philosophers argue that rational human beings meriting respect also have concomitant duties in relation to other members of the society. Locke expressed this in his statement that "no one ought to harm another in his life, health, liberty, or possession."[16] To assist the patient in reaching a self-protective state so that he or she can fulfill this social duty is also a respectful intervention and a goal that undergirds many state commitment laws. For example, California and Wisconsin, as well as other states, allow for involuntary commitment and treatment of individuals who are imminently dangerous to themselves or others. Such states also provide for involuntary treatment of individuals who are gravely disabled to the extent of being unable to provide for their own food, clothing, or shelter. Such legislation promotes the goal of treating these individuals to restore them to a state of self-protection.

In sum, this chapter argues that the commitment of a dangerous mentally ill individual to a humane, sensitive, and effective treatment program can be a most ethical action, providing extreme respect for the individual and his potential to return to a self-protecting state. However, when the situations of gravely disabled or dangerous behaviors are not present, society has chosen to protect the refusing but innocuous patient from mandatory therapy.

Correlative Elements in Responsible Treatment

There are correlates to the proffering of respectful treatment to patients, particularly objecting patients. Legal discussion of such issues includes cases such as *Donaldson* v. *O'Connor*.[17] If the rationale for treatment, even in the midst of objection, is to restore the person to a more self-protective state, then this must be possible in the instituted treatment. Mere custodial care does not meet this social duty. The benefit for the patient is precisely the restoration of the human attributes of self-protection and self-regulation. Psychiatrist-made decisions that either reduce patients to mere object dependence or take advantage of the present defenselessness of the patient are contrary to the principle of respect.

This responsibility of treatment must extend beyond the early stages of intervention. Interventions by psychiatrists, social engineers, and politicians in the late 1960s and early 1970s in California moved state-hospital patients out into the community with the goals of enhancing self-dependency and reducing institutionalization. The reality of this transition was the development of halfway houses where chronic psychiatric patients failed to receive any treatment or support in independent living, where they were at times locked out during the day to roam the city streets or public square, and where, politically, they became fiscal liabilities, objects to be manipulated in the state budget, rather than humans toward whom society was directing a respectful, caring intervention.[18]

The psychiatrist should be fully cognizant of the fact that treatment and behavior modification may entail more than the administration of a prolixin injection every two weeks. Treatment may involve active social action to provide adequate living environments for the patient. Treatment may require the initiation of social-skills courses so that the patient will become able to live independently, manage a budget, shop for groceries and clothes, and learn how to utilize the bus routes in the city. Failure to provide these elements in the treatment process undermines the goal of restoring this person to a truly self-protective, independent state and therefore fails to afford a full level of respect for the patient undergoing treatment. Such failure, in fact, denies the original ethical underpinnings for the decision to treat.

Similarities between Education and Psychiatric Behavior Modification

A number of similarities exist between the characteristics of behavior modification in formal education and the pharmacologic and behavioral interventions of psychiatry. John Dewey speaks of the objectives of instruction and discipline in education as being to:

> Prepare the young for future responsibilities and for success in life, by means of acquisition of the organized bodies of information and prepared forms of skill which comprehend the material of instruction.[19]

Huxley refers to education as:

> Instruction of the intellect . . . and the fashioning of the affections and of the will into an earnest and loving desire to move in harmony with laws (of nature, men and their ways).[20] . . . [and] to make yourself do the

> thing you have to do, when it ought to be done, whether you like it or not.[21]

These definitions or goals of education are strikingly similar to the psychotherapeutic goal to restore the person to a state of self-protection wherein the individual can "devise plans, pursue them, modify them, engage the interest and win the support of others for them—all against the adversity of circumstances, the threats of environment, and the self-interest of others . . ." as well as to "mobilize one's inner resources and outer circumstances to attain a goal which others will acknowledge."[22] An identical goal for both education and therapy is to increase the freedom of the individual who is being educated or treated (in both cases, exposed to behavior modification) by reducing his vulnerability in living through the development of more functional perceptions of his or her world and by the acquisition of tools and techniques to manipulate the environment toward his or her own ends. Dewey points out that:

> Since freedom resides in the operations of intelligent observation and judgement by which a purpose is developed, guidance given by the teacher to the exercise of the pupils' intelligence is an aid to freedom, not a restriction upon it.[23]

Nonetheless, for both education and psychiatric treatment, there exists a tension between the concept of total personal freedom and an attempt *in loco parentis* to facilitate the creation of a self-protective state within an individual.

Another similarity between education and psychiatric behavior modification is the incomplete environmental control of the intervention process. Children in our society experience formal education during only a portion of their childhood hours; other behaviors involving family activities or, in our culture, television viewing account for many more hours than those involved in the formal educative process. This is also true for nearly all of the psychiatric behavior modification methods in use. Therapy interventions may entail one hour per week. Even for the more disturbed patient in a day-hospital setting or on an inpatient ward, treatment geared to behavior modification will occur in less than half of the person's waking hours. In spite of fictional novels such as *Walden Two,* the fact remains that psychiatric interventions of behavior modification are not all encompassing.[24] Exceptions might be certain token economy systems that restrict a small number of patients to many tightly designed contingency patterns of reinforcement.[25] Most therapy modalities in mental health treatment occur in a milieu more closely approximating that of formal education than that found in this country's penal system, in which environmental control is nearly complete.

Nonpsychiatric Social-Engineering Issues

Some of the issues in behavior modification highlighted by such novels as *A Clockwork Orange* involve social, not medical, issues.[26] Social engineering engenders a tension between personal freedom and social conformity, between the individual's autonomy and his subservience to the social good as perceived by a society's leaders. Our society permits considerable social deviance in contrast with the People's Republic of China. Society's acceptance of deviance and the manipulation of behavior to reduce deviance are not psychiatric problems; they are social problems.

At the present time, in relation to criminal social deviance, our society has generally chosen to punish such behavior as a deterrent. However, the nations' courts have refused to permit certain types of behavior modification techniques within the prison system, even if these would effectively rehabilitate the prisoner and assure a decrease in socially unacceptable behavior. For example, in *Clonce* v. *Richardson,*[27] the courts halted a behavior modification program that involved sensory and motor restrictions, stating that conditions of initial incarceration with these environmental restrictions required that due process was required for placement of individuals into these exceptional conditions, a ruling that effectively brought the program to a halt. In brief, the courts currently maintain a posture that the prisoner may refuse treatment, the goal of which is to reduce socially punishable deviant behavior.

Certain legal theoreticians have proposed alternatives to a legal system based upon the premise of punishing the deviant person. Lady Wootton, an English parliamentarian, has argued that, following the conviction of an individual, he or she should be sentenced to appropriate treatment that will lead to a reduced probability of the recurrence of the socially unacceptable behavior.[28] An alteration of our society's penal system in the direction of such a plan of rehabilitation is a social, not a medical, decision. The modification of behavior by either pharmacological or behavioral intervention techniques in such a rehabilitative system is not a medical problem. Drug giving and behavior modification in such a system may be roles from which mental health workers and psychiatrists in particular may wish to abstain.

Conclusion

This chapter addressed aspects of psychiatry's ability to modify behavior by behavioral and pharmacologic techniques. It pointed out that many of these modification techniques are ones of modulation rather than tight control. However, the fact that these tools do change behavior opens the issue of their ethical application.

The principle of respect for persons is an ethical principle that can assist immensely in the ethical application of psychiatric interventions. Central to the ethical application of behavior modifying techniques is the goal to restore the individual to a self-protective state.

Although freedom is greatly prized in our society and the affording of freedom to an individual is a means of providing him with respect, this chapter argued that in situations of imminent danger to self or others it is more respectful to intervene, even against an individual's wishes, than it is to allow him to be used as an object, valuing freedom above respect for persons.

Most forms of psychiatric intervention and behavior modification have much in common with the social institution of education. During the process of modification, both systems limit freedom, yet both are working to produce a greater degree of personal freedom, environmental control, and lifestyle options following the modification process. Also, most formal education and nearly all psychotherapy occur during only a fraction of the subject's waking hours and usually in a relatively open environment.

More restrictive environments have been proposed for behavior modification within the prison system. This is a social issue rather than a medical/psychiatric issue. The degree to which social conformity is to be engineered and the types of deviance that will and will not be permitted are social questions, not medical questions. Should a society choose to institute formal behavior modification techniques to decrease deviance, no a priori reason exists why health professionals should be involved in this process.

Notes

1. P. London, *Behavior Control* (New York: Harper & Row, 1979).

2. J.M. Davis and J.O. Cole, "Organic Therapies," in *Comprehensive Textbook of Psychiatry,* vol. 2, edited by A.M. Freedman; H.I. Kaplan; and B.J. Sadock, chapter 31 (Baltimore: Williams and Wilkins Company, 1975).

3. P.A. Berger, "Antidepressant Medications and the Treatment of Depressions," in *Psychopharmacology—From Theory to Practice,* edited by J.D. Barchas; P.A. Berger; R.D. Ciaranelto; and G.R. Elliott, pp. 174–207 (New York: Oxford University Press, 1977); and APA, *Electroconvulsive Therapy,* Task Force Report no. 14 (Washington, D.C., 1978).

4. S. Geishon and B. Shopsin, eds., *Lithium: Its Role in Psychiatric Research and Treatment* (New York: Plenum Press, 1973).

5. T. Ayllon and N.H. Azrin, *The Token Economy: A Motivational System for Therapy and Rehabilitation* (New York: Appleton-Century-Crofts, 1968).

6. J.B. Watson, *Behaviorism* (Chicago: University of Chicago Press, 1924), p. 104.

7. F. Redlich and R.F. Mollica, "Overview of Ethical Issues in Contemporary Psychiatry," *American Journal of Psychiatry* 133 (1976):135.

8. B. Eichelman and J. Barchas, "Ethical Aspects of Psychiatry," in *Psychiatry and Behavioral Sciences,* edited by D.A. Hamburg; J.D. Barchas; P.A. Berger; and G.R. Elliott (New York: Oxford University Press, in press).

9. H. Thoreau, *Civil Disobedience* (Boston: D.R. Godine, 1969).

10. I. Kant, *Groundwork of the Metaphysic of Morals,* transcribed and analyzed by H.J. Paton (New York: Harper & Row, 1964).

11. E.E. Harris, "Respect for Persons," in *Ethics and Society,* edited by R.T. DeGeorge, p. 113 (Garden City, N.Y.: Anchor Doubleday, 1966).

12. Redlich and Mollica, "Overview of Ethical Issues."

13. J.P. Brady, "Systematic Desensitization," in *Behavior Modification: Principles and Clinical Applications,* edited by W.S. Agras. (Boston: Little, Brown & Company, 1972).

14. A. Jonsen and B. Eichelman, "Ethical Issues in Psychopharmacologic Treatment," in *Legal and Ethical Issues in Human Research Treatment: Psychopharmacologic Considerations,* edited by D.M. Gallant and R. Force (Jamaica, N.Y.: Spectrum Publications, 1978), p. 149.

15. Ibid.

16. J. Locke, *Second Treatise on Human Nature,* edited by P. Laslette (New York: Mentor Books, 1965), p. 311.

17. Donaldson v. O'Connor, 493 F.2d 507 (5th Circ. 1974).

18. A. Davis, S. Dinitz, and Benjamin Pasamanick, *Schizophrenics in the New Custodial Community* (Columbus: Ohio State University, 1974).

19. J. Dewey, *Experience and Education* (New York: Collier Books, 1963), p. 18.

20. T.H. Huxley, "A Liberal Education," in *Autobiography and Selected Essays,* edited by Ada Snell, p. 40 (Boston: Houghton Mifflin Co., 1909).

21. T.H. Huxley, *Technical Education and Other Essays* (New York: Humbolt Publishing, 1885), p. 5.

22. Jonsen and Eichelman, "Ethical Issues."

23. Dewey, *Experience and Education.*

24. B.F. Skinner, *Walden Two* (New York: Macmillan, 1962).

25. Ayllon and Azrin, *The Token Economy.*

26. A. Burgess, *A Clockwork Orange.* (New York: W.W. Norton, 1963).

27. Clonce v. Richardson, no. 73 CV 373-S (W.O. Mo., 31 July 1974).

28. B. Wootton, *Crime and Criminal Law* (London: Stevens Publishing Company, 1963).

12 The Rights of Methadone-Maintenance Patients

Andrew A. Sorensen

One of the most popular arenas in which the rights of patients have been debated is that of the treatment of drug addicts. Suggestions have ranged from increasing the severity of sentences for those who sell and use addicting drugs to decriminalizing addiction and regarding it as a disease or disorder that requires medical treatment. In the latter camp are those who favor treating persons addicted to any of the opiates with methadone, acknowledging that many of these persons will be maintained on methadone for many years.

It is appropriate that a discussion of the ethical issues attendant on the use of methadone begin with a necessarily truncated historical overview of the use of addictive drugs in our society. This is so because the responses of our society to the issues of freedom of choice and the rights of patients in methadone-maintenance programs are conditioned, in large measure, by the evolution of our societal use of addictive drugs and our collective reactions over time to the treatment of addiction.

Historical Overview of Drug Addiction

Even before the establishment of our republic, concern had been expressed about the use of addicting drugs. Ben Franklin reported in his *Gazette* a "surprising tho' authentick" account that the total of rum imported into Philadelphia in 1728 ran to 2,124,500 gallons. With such enormous amounts of alcohol being imported, it is no wonder that reports of drunkenness became more frequent. Benjamin Douglass, a resident of Newport, came to an ignominious end in 1728 when, "being in drink and going into a little house [he] fell down from ye seat and broke his neck." In 1741 a child in Philadelphia, "about 4 or 5 years of age, died by drinking a large quantity of rum." In order to curb these excesses, many communities enacted laws against immoderate drinking. Boston, for example, posted lists of drunkards, "as the Law Directs," and moralists began to flood news-

I acknowledge the helpful comments of Colleen Clements, Burr Eichelman, and Michael Weintraub on an earlier draft of this chapter and Helen Weeks for typing the manuscript.

papers and pamphlets with the claim the "abuse of Strong Drink is becoming Epidemical." Yet neither the laws nor the warnings seemed to have much effect on the intake of beverage alcohol. In 1792, the annual per capita consumption was 2.5 gallons. By 1810 that figure had almost doubled, to 4.7 gallons, and by 1823 it had increased to 7.5 gallons. This represents a trebling of the per capita consumption in three decades.[1] There is little question that the geometric increase in the consumption of and the problems attendant to what we now call alcoholism contributed to the emergence of the National Prohibition Party in 1869 and the Women's Christian Temperance Union in 1874.[2]

After the Civil War, Americans were also being warned with greater frequency about the spread of opium addiction. A study of 235 "habitual opium eaters" in Chicago in 1880 ominously reported that 45 percent were housewives while only 3 percent were "society ladies" and that three times as many women as men were addicted.[3] Five years later a study was done of opium addiction in Iowa. In this study of 235 habitués, 86 were males (including 18 physicians) and 129 were females, although we are not told how many were society ladies.[4] That opium addiction had also spread to the Appalachian hills was indicated by a Kentucky physician who, in 1878 and 1880, described the effectiveness of cocaine as an antidote to opium addiction.[5] Parenthetically, it is interesting to note that Sigmund Freud cited a reference to the latter article in his earliest paper on cocaine.[6] The disparity of signs and symptoms for which opium was prescribed disturbed many nineteenth-century observers of these phenomena. Virgil Eaton, writing in the *Popular Science Monthly* in September 1880, reported:

> I was surprised to learn how extensively opium and its alkaloids—particularly sulphate of morphia—are used by physicians. I found them prescribed for every ailment which flesh is heir to. They are used for headache, sore eyes, toothache, sore throat, laryngitis, diphtheria, bronchitis, congestion, pneumonia, consumption, gastritis, liver-complaint, stone in the gall-duct, carditis, aneurism, hypertrophy, peritonitis, calculus, kidney trouble, rheumatism, neuralgia, and all general or special maladies of the body. It is the great panacea and cure-all.[7]

Eaton was not content to convey anecdotes and journalistic impressions, however. As with those who conducted the studies in Chicago and Iowa, Eaton decided to get a quantitative assessment of prescribing practices. He visited pharmacists in Boston and randomly selected prescriptions (or recipes, as they were then called):

> [In] the first [drug] store I visited I found 42 recipes which contained morphine out of the 300 examined. Close by, a smaller store, patronized by poorer people, had 36. Up in the aristocratic quarters, where the customers call in carriages, I found 49 morphine recipes in looking over 300.

> At the North End, among the poor Italian laborers, the lowest proportion of 32 in 300 was discovered. Without detailing all the places visited, I will summarize by saying that, in 10,200 recipes taken in 34 drug stores, I found 1,481 recipes which prescribed some preparation of opium on an average of fourteen and one half percent of the whole.[8]

An editor of the *Journal of the American Medical Association (JAMA)* took umbrage at Eaton's attribution of physician responsibility for the spread of addiction through careless prescribing. The editor suggested that, had Eaton considered the even more widespread use of patent medicine, "whose popularity depends on the amount of alcohol or morphine they contain," he would have presented quite a different picture.[9] The *JAMA* editor's comments obviously demonstrate that he did not anticipate the addictive use of opiates by physicians themselves. For example, Sir William Osler reported that Dr. William Stewart Halstead, one of the four distinguished founders of the Johns Hopkins Medical School, had "cured his cocaine habit by turning to morphine."[10]

After the turn of the twentieth century, a diverse consortium supported the demands of muckrakers and reformers to regulate the use of nonalcoholic addictive drugs quite apart from the efforts of the temperance movement. Especially the pharmacists sought to protect their prestige and called for physicians to be more discriminating in their prescribing practices. Drug abuse was often linked with crime and instability. Some social workers argued that addiction "fostered indolence and poverty, threatened family ties, and inhibited individual development."[11] The escalation of concern about the use of opiates led to the passage of the Harrison Act in 1914, which implicitly institutionalized a medical solution to the problem of addiction, although it also increased governmental regulation of physician behavior—addicts could legally obtain opiates only through physicians as part of a cure regimen, and institutional treatment programs were established under medical supervision. This set of events was in marked contrast to what was happening in Great Britain. Although the Home Office there had wanted to monitor physicians' practices more closely, the Rolleston Report of 1926 "legitimized the doctor's right to maintain his addict on drugs" without the restrictions his U.S. counterpart experienced.[12]

One of the earliest U.S. proponents of the medical treatment of drug addiction was Dr. Charles Terry, who set up a narcotic clinic in Jacksonville, Florida, in 1912.[13] Following the armistice ending World War I, many heroin-maintenance programs were begun throughout the country. In part, the establishment of these clinics was stimulated by two Supreme Court decisions handed down in 1919 that had determined that "the Harrison Act was constitutional and that doctors who maintained addicts were in violation of the Harrison Act."[14] As agents of the Narcotics Bureau of the Internal Revenue Service began serving indictments against physicians who were treating addicts on an outpatient basis, the population of untreated addicts was perceived to be growing dramatically, with the veterans coming home from the war accounting for much of the increase. In a remarkably short period, clinics sprang up to treat addicts. Two of the principal reasons given then for mounting these programs were that addiction to opiates had reached epidemic proportions; and it was widely conceded that once a person was addicted to a drug, few treatments had proved effective in detoxifying the addict. So until more was known about the effectiveness of detoxification, and/or until the addicts were willing to quit using drugs, addiction regulated by the federal grovernment—it was averred—would lead to reduced crime and increased family stability and therefore was to be preferred to unregulated addiction. Consequently, in the years between 1910 and 1920 many heroin- and morphine-maintenance clinics were opened in which the behavior of physicians and addicts could be more closely monitored.[15]

Although the mechanism of distribution of the opiates had been altered dramatically, the medicalization of addiction was still the order of the day. Thus, the administrator of Narcotic Drug Control in the metropolitan New York area wrote in 1921: "It is at once evident that a very important power is given to the physician and that a great deal depends on the use he makes of this power."[16] This power had been compromised by the passage of the Volstead Act, however, which gave the Prohibition Unit of the Internal Revenue Service the responsibility for enforcing the Harrison Act. It was soon evident that the prohibition ideology was being extended to other addictive drugs. For several reasons—among them the expressed feeling that the patients enrolled in such programs were abusing them and the aggressive enforcement of the Harrison Act—the morphine- and heroin-maintenance programs were disbanded. We then entered an era of increasing criminalization of addiction in which it was felt that addiction could be virtually eliminated by tougher laws and more rigorous enforcement of those laws. That mentality was epitomized by the appointment in 1930 of Harry Anslinger as the commissioner of the Bureau of Federal Narcotics.[17]

Support for opiate-maintenance programs lay dormant until 1964—two years after Anslinger's retirement. The then husband-and-wife team of physicians Vincent Dole and Marie Nyswander announced that a heroin derivative named methadone, discovered in 1938 by German scientists who were searching for a synthetic equivalent to morphine for the Nazi field hospitals, could be used as a narcotic substitute for heroin in the management of addiction.[18] Two of the arguments for setting up methadone-maintenance clinics were remarkably similar to those for heroin and morphine maintenance a half century earlier. First, it was felt that addiction was reaching epidemic proportions, that "the drug explosion of the sixties led to a vast increase in the number of heroin addicts,"[19] and that treatment programs were needed to respond to the drug epidemic. Second, because heroin-maintenance programs were thought to be medically ineffective and/or politically undesirable, an alternative was needed that would reduce crime and dependence on welfare through increased employment while promoting family stability. Thus, it was contended, enormous social and economic costs would be saved. Additionally, arguments for the pharmacological benefits of methadone maintenance were cited.[20]

As this cursory overview indicates, throughout the history of drug use in our society, we have distinguished between licit and illicit drugs.[21] However, these distinctions have not been characterized by logical consistency. At some times it has been the case that drugs have been regarded as licit in some circumstances and illicit in others. At other times certain drugs have been considered illicit in any circumstance, and other drugs have been considered licit in any circumstance. As pharmacologist Robert Byck so trenchantly observed, "We choose our poisons on the basis of tradition, not pharmacology."[22]

Ethical Problems Raised by Societal Attitudes toward Addiction

From the time when Puritan divines were a dominant influence in New England until now, fulminations against the evils of addiction have been accorded a mixed reception. One of the most consistently prolix among the many issues in the history of drug use has been the formulation of norms by societies to regulate such use. In most societies, substantial ambivalence exists regarding drug use. In a few societies such as the Islamic Republic in Iran, the use of some addicting drugs such as alcohol and opium is forbidden (although Avicenna, perhaps the most famous of all Islamic physicians, is said to have died in 1037 from an overdose of opium,[23] while the use of other addicting drugs such as nicotine is tolerated.[24] Other societies such as ours have decreed the nonmedical use of some addicting drugs to be illegal in all circumstances (for example, opium and heroin), while the nonmedical use of other addicting drugs, in prescribed circumstances, is legal (for example, alcohol). As is well known, our cultural ambivalence is also extended to nonaddictive drugs. Thus, in most areas of our country, the use of marijuana is illegal, but the use of caffeine and nicotine is legal. Also, the government spends millions of dollars each year to suppress the marijuana industry, while it spends millions more to subsidize the tobacco industry.

In support of the decriminalization of the use of marijuana, many people have argued that citizens have the natural right to use whatever drug they choose. However, it is clear that they do not have the man-given—

that is, social and legal—right to do so. People under the age of 18 in some states and 21 in others, for example, cannot legally drink alcohol, even if it is served to them by their parents in their own homes. However, people under 18 may legally use methadone as long as they demonstrate that they are addicted to it and that the methadone they use is distributed in federally regulated programs. This set of events, which is quite bizarre to say the least, has been instigated largely through the efforts of police and judicial officers and has received substantial assistance from the medical profession. The collaborative efforts of the medical and legal establishments have led to a situation in which citizens are seldom legally able to choose the use of an addictive drug, except in circumstances dictated by the government.

This restriction of the freedom of choice of consumption of drugs certainly cannot be defended rationally by arguing that the goods at issue need to be regulated because they cause harm and/or because they have addictive properties. For example, the chronic long-term use of alcohol will cause greater harm than the similar use of many narcotics.[25,26] Also, obese people frequently testify that they feel addicted to food.[27] Yet the government spends infinitely less to regulate the sale and use of alcohol than narcotics and has no organized effort to regulate the quantity of consumption of food, although obesity is correlated with many health problems and results in great social costs.

Another manifestation of our ambivalence is that not only has our society regulated the consumption of certain drugs, but also it has provided threats and inducements for the treatment of addiction to certain drugs, but not to others, by chemical modification of behavior. In many states, persons convicted of the use of some addictive drugs are offered placement in methadone-maintenance programs in lieu of incarceration—frequently through the mechanism of plea bargaining. It is ironic to note that this is done in the name of enlightenment—that is, the chemical modification of the behavior of addicts is regarded as more humane and progressive than imprisonment. If society has punished the individual for choosing heavy use of a drug that the society has defined as illegal when taken without medical supervision, then one must ask if using precisely the same drug in equivalent amounts under medical supervision, when chosen as a lesser evil than incarceration, is, in fact, a greater good.

Many problems are created by the medicalization of addiction and the lifelong stigma it conveys. For one thing, intentionally or unintentionally, it engenders a dependence of the patient on the physician that is usually initiated by neither party. One of my physician colleagues has argued that there is a social sanctity to the healing function in the usual physician-patient relationship that is abrogated when the patient involuntarily comes to the physician for treatment, not because the patient feels he has a medical problem but because the society has dictated to the patient that he must seek medical treatment. My colleague does not argue that this quality is peculiar to instances of mandated treatment of drug addiction, but he feels that these circumstances "weaken the traditional patient-physician relationship."[28] A second problem with the medicalization of addiction is that the patients often seem to be desensitized to the need for social change and singularly uninvolved in political activities. Yet another problem is that males who are on methadone for sustained periods have been found to have low sperm counts and to be near sterility.[29]

There are also problems with dependence on the law-enforcement profession. Among these problems is the increased probability that, in cases involving the use and/or sale of illicit drugs, addicts who have been paroled or discharged from prison are often considered suspects. Another problem is the lifelong stigma that conviction on a drug charge conveys to prospective employers.

Given these ramifications of addiction, it seems that neither the medical nor nonmedical disposition of addiction is inherently superior to the other. Users of narcotics who are arrested and convicted for such use are ostensibly given a choice between medical treatment and imprisonment. However, many addicts—because of the fear of the sequelae of imprisonment or their ability to detoxify cold turkey and/or the uncertainty that they will be able to secure illicit drugs in prison—do not consider prison a tenable option. This bifurcation of alternatives between medically supervised chemical modification of behavior or abstinence from drugs in prison severely limits the right of addicts who are presented only these two modalities to refuse medical treatment for addiction, although it expresses society's perceived obligation to provide treatment, and surely reinforces the notion that addiction is both deviant and a disease.

Two issues here concern me profoundly. The first is that the presentation of this option violates the principle of respect. Jonsen and Eichelman have eloquently articulated their concern regarding the physician's respect for the patient:

> Acting for the benefit of the patient necessarily involves an assessment of the values represented by the concerns of others in the patient's community. This assessment should be aimed at revealing what is necessary for the patient to return to a state of self-regulation and self-protection. The benefit to the patient is precisely the restoration of these human attributes. Decisions made by psychiatrists which either reduce patients to more abject dependence or take advantage of the present defenselessness of the patient are contrary to the principle of respect.[30]

In methadone-maintenance programs, in which the responsibility for the chemical regulation of the patient is exercised by the medical profession exclusively rather than shared between the patient and the physician, the principle of respect is routinely ignored.

The second issue of concern is whether the societal sense of obligation for the perceived good of the citizenry should dominate over the will of the respective citizens to have their rights guaranteed. For too long we have looked to medical technology to help us make our decisions regarding the treatment of addiction. Many proponents of behavior modification fall into this camp. Instead of addressing the thorny issue of the balance between community obligations and individual rights, we have subscribed, without sufficient debate, to the notion that the social contract implies that society has the right to modify and intervene in individual actions. Thus we have chosen physician technicians and a gaggle of legislators and enforcers of the law to provide the answers to the problems of addiction that face us. The result has been what Dorothy Nelkin describes as a technological fix—that is, we circumvent social problems by reducing them to technological problems.[31] For millennia, addiction—although not labeled as such—has been defined as a social problem. However, if one concurs in the observation that, especially latterly, it has been thought of as a disease, the societal expectation that the medical profession can discover a solution to the problem seems to have been elevated. The role of the physician in this denouement has been transformed from that of healer to that of therapist cum social engineer.

One of the most vivid examples of such role transformation in recent years is the development of LAAM.[32] It was reasoned that if the once-daily administration of methadone, which is supervised by physicians, was medically adequate but required too much manpower and too many and/or too complex demands on the physician-patient relationship, then an equivalent drug that had essentially the same effects but that required less manpower would be even more desirable. A team of physicians, evaluating a methadone-maintenance program, described it as a "technically simple and low-cost activity."[33] Yet, if this is so for daily administration of methadone hydrochloride, how much simpler and cheaper, it was argued, is LAAM—which, because of its longer half-life, needs to be administered only three times weekly? A study of the efficacy of LAAM, also organized by a team of health care providers, notes that using LAAM permits the patients to avoid "demanding daily clinic attendance, which imposes hardships on patients."[34]

The advocacy of switching from methadone hydrochloride to LAAM may be viewed as fostering less reliance on medical professionals because it reduces the amount of time the addicts spend in the health care system. Viewed from another perspective, however, the development of LAAM may be seen as encouraging even greater dependence on medical technology. If LAAM can be administered only once every three days, then perhaps—one might hope—the technology will be able to provide a drug that will be effective for a week or a month.

About three years before the Watergate burglary was executed, then President Richard Nixon had the temerity to proclaim addictive drugs "Public Enemy No. One."[35] Shortly thereafter, he established the Special Action Office of Drug Abuse Policy (SAODAP), whose "chief commitment was to the rapid expansion of methadone maintenance."[36] At that time I was employed as an epidemiologist in a methadone-maintenance program in which an experience I had made me feel as if I were exploring Huxley's *Brave New World*.[37] We received a notice from SAODAP in 1971 informing us that they were exploring the concept of implanting a narcotic antagonist in nonaddicted children who were considered at high risk for subsequent addiction that would give them a lifelong blockade of the euphoric effects of opiates. This, I submit, is the technological fix gone wild.

Recommendations for Treatment of Addiction

Our current state of affairs with respect to the treatment of drug addiction is in need of massive reform, but the situation is not hopeless. Therefore, I should like to conclude with three recommendations for the modification of our societal approach to the treatment of addicts.

First, we ought to affirm the right of people to be addicted, providing that the addiction does not cause harm to others and that it does not make inordinate demands on our resources. This does not necessarily mean that one supports the increased or decreased incidence of prevalence of drug addiction. In *On Liberty,* John Stuart Mill wrote

> The only purpose for which power can be rightfully exercised over any member of a civilized community against his will is to prevent harm to others. His own good, either physical or moral, is not a sufficient warrant. He cannot rightfully be compelled to do or forbear because it will make him happier, because in the opinions of others, to do so would be wise or even right.[38]

However, several centuries of experience in an astonishingly wide array of cultures have indicted that we have not been at all successful in prohibiting addiction. Also, there is no unequivocal evidence that intensifying the severity of punishment for addiction and/or medicalizing it by regarding it as a disease in and of itself has had any effect on reducing the incidence and prevalence of addiction.

Second, just as people have the right to be addicted, so they have the right not to be addicted. Evidence has been presented that the quantity of alcohol consumed is proportional to its availability, although there is some ambiguity in these reports.[39] Nevertheless, alcoholism is positively correlated with the amount of alcohol individuals consume. Thus it seems that while restricting the availability of addictive drugs will not eradicate addiction, it may reduce addiction or inhibit growth in the incidence and prevalence rates of addiction. Conversely, one has every reason to hypothesize, a priori, that increasing the availability of addictive drugs will be positively correlated with an increase in the proportion of addicts in the population to a given plateau.

Thus, the right of citizens to be addicted must be weighed against the right of citizens not to be addicted. To appeal again to Mill's dictum, in order to prevent harm to others, we may not be able to totally withdraw all restrictions on the availability of addictive drugs. Therefore, it seems that we should reject this sentiment expressed by Thomas Szasz: "I favor free trade in drugs for the same reason the founding fathers favored free trade in ideas. In an open society, it is none of the government's business what idea a man puts into his mind; likewise, it should be none of the government's business what drug he puts into his body.[40] By steering between the Scylla of what Szasz calls the ethic of individualism, and the Charybdis of collectively regulated behavior, we can approach the formulation of restrictions on the use of addictive drugs more rationally than we have in the past.

Finally, we must recognize that the curious marriage between increased medicalization and the concomitant criminalization of drug addiction is not ethically neutral. The values reflected in these social trends should be remarked and examined in light of the effects they have on the rights of citizens to choose addiction and, if arrested and convicted for exercising that right, to choose among the modes of treatment for that addiction. As it stands now, all too often for persons who are illegally addicted to opiates, arrest and conviction results in what is ostensibly a voluntary choice between incarceration in an environment that is supposedly free of addictive drugs and chemical modification of behavior by an addictive drug in a medically supervised setting. In some circumstances, a third alternative is presented—namely, the convicted person may choose to participate in a drug-free therapeutic program outside of prison.

What I am pleading for, among other things, is increasing the number of therapeutic choices for those citizens whose addiction makes them liable to arrest. Here are just a few additional alternatives: (1) heroin-maintenance programs modeled after the system used in the United Kingdom;[41] (2) oral opium maintenance; (3) using difenoximide[42] or propoxyphene napsylate in detoxification from heroin[43] or methadone;[44] (4) work-rehabilitation programs; and (5) abstinence-oriented therapeutic communities of various sorts.[45] Widening the range of alternatives would enhance the probability that informed consent would be elevated from the level that Eichelman (chapter 11) has called passive consent to that of active consent.

Although the choices presented to addicts regarding enrollment in methadone-maintenance programs are inadequate, the fault for this limitation does not lie with Dole and Nyswander. As Zinberg so correctly noted, their "prescience about the problems of methadone maintenance has been remarkable."[46] This prescience was manifested in the concluding statements of their ten-year perspective on methadone-maintenance treatment: "As in other areas of medical practice, the question is not how many 'treatment slots' are available . . . but what quality of treatment the patients receive. Bureaucratic control of methadone programs has given us 'slots,' a rule book, and an army of inspectors, but relatively little rehabilitation."[47]

One would hope that increasing the choices for persons convicted of drug addiction and exhibiting greater concern for their rights would enhance the likelihood that rehabilitation could occur.

Notes

1. Paul C. Conley and Andrew A. Sorensen, *The Staggering Steeple: The Story of Alcoholism and the Churches* (Philadelphia: Pilgrim Press, 1971).

2. Ibid.

3. Charles W. Earle, "The Opium Habit: A Statistical and Clinical Lecture," *Chicago Medical Review* 2 (1880):442–446.

4. J.M. Hull, "The Opium Habit," in *Third Biennial Report,* edited by Iowa State Board of Health, pp. 535–545 (Des Moines: George E. Roberts, 1885).

5. W.H. Bentley, "Erythroxylon Coca," *Therapeutic Gazette* 1 (June 15, 1980):350–351. See also Bentley, "Erythroxylon Coca in the Opium and Alcohol Habits," *Therapeutic Gazette* 1 (September 1880):253–254.

6. Sigmund Freud, "Uber Coca," *Centralblatt fur die gesammte Therapie* 2 (1884):289–314.

7. Virgil G. Eaton, "How the Opium Habit Is Acquired," *Popular Science Monthly* 33 (1888):663–667.

8. Ibid.

9. Editorial, "How the Opium Habit Is Acquired," *Journal of the American Medical Association* 11 (1888):419–420.

10. Edward M. Brecher, and the editors of *Consumer Report, Licit and Illicit Drugs: The Consumers Union Report on Narcotics, Stimulants, Depressants, Inhalants, Hallucinogens, and Marijuana—Including Caffeine, Nicotine and Alcohol* (Boston: Little Brown & Company, 1972), pp. 33–34.

11. H. Wayne Morgan, *Yesterday's Addicts: American Society and Drug Abuse 1865–1920* (Norman: University of Oklahoma Press, 1974), p. 180.

12. "Report of Ministry of Health Departmental Committee on Morphine and Heroin Addiction." H.M. Stationery Office, 1926.

13. Charles Terry and Mildred Pellens, *The Opium Problem* (Montclair, N.J.: Patterson Smith, 1928).

14. Don Waldorf; Martin Orlick; and Craig Reinarman, *Morphine Maintenance: The Shreveport Clinic, 1919–1923* (Washington, D.C.: Drug Abuse Council, 1974), pp. 2–3.

15. David F. Musto, *The American Disease: Origins of Narcotic Control* (New Haven: Yale University Press, 1973), pp. 151–182.

16. Sara Graham-Mulhall, "Experiences in Narcotic Drug Control in the State of New York," *New York Medical Journal* 113 (1921):106–111.

17. Musto, *The American Disease,* pp. 210–214.

18. Vincent Dole and Marie Nyswander, "A Medical Treatment for Diacetylmorphine (Heroin) Addiction: A Clinical Trial with Methadone Hydrochloride," *Journal of the American Medical Association* 193 (1965):646–650.

19. Norman E. Zinberg, "The Crisis in Methadone Maintenance," *New England Journal of Medicine* 296 (1977):1000–1002.

20. Dole and Nyswander, "A Medical Treatment."

21. Edward M. Brecher, and *Consumer Reports, Licit and Illicit Drugs.*

22. Robert Byck, ed., *Cocaine Papers by Sigmund Freud* (New York: Stonehill Publishing Co., 1975).

23. D.I. Macht, "The History of Opium and Some of its Preparations and Alkaloids," *Journal of the American Medical Association* 64 (1915):477–481.

24. I am defining addictive drugs in the same sense as Jerome H. Jaffe, "Drug Addiction and Drug Abuse," in *The Pharmacological Basis of Therapeutics* edited by L.S. Goodman and A. Gilman, (New York: Macmillan, 1975), pp. 285–305.

25. Ernest Saward and Andrew Sorensen, "The Current Emphasis on Preventive Medicine," *Science* 200 (1978):889–894.

26. Edward M. Brecher, and *Consumer Reports, Licit and Illicit Drugs.*

27. David A. Levitsky, "The Fat Man and the Addict: The Biology of Addiction," *Human Ecology Forum* 3 (1973):2–4.

28. Burr Eichelman, letter to Andrew A. Sorensen, 15 June 1979.

29. W.F. Wieland, "Sexual Effects and Side Effects of Heroin and Methadone," in *Proceedings of Third National Conference on Methadone Treatment* (Washington, D.C.: U.S. Government Printing Office, 1970), pp. 50–52.

30. Albert R. Jonsen and Burr Eichelman, "Ethical Issues in Psychopharmacologic Treatment," in *Legal and Ethical Issues in Human Research and Treatment,* edited by Donald M. Gallant and Robert Force, pp. 143–174 (New York: Spectrum Medical and Scientific Books, 1978).

31. Dorothy Nelkin, *Methadone Maintenance: A Technological Fix* (New York: George Braziller, 1973).

32. Jack D. Baline; Pierre Renault; Gail L. Levine; and John A. Whysner, "Clinical Use of LAAM," in *Recent Developments in Chemotherapy of Narcotic Addiction,* edited by B. Kissin; J. Lowinson; and R. Millmar, pp. 214–231 (New York: New York Academy of Sciences, 1978).

33. William Cleveland; Brian Bowles; William Hicks; Charles Burks; and Kenneth Rogers, "Outcomes of Methadone Treatment of 300 Innercity Addicts," *Public Health Reports* 89 (1974):563–568.

34. Walter Ling; Charles Charuvastra; Samuel C. Kaim; and James Klett, "Methadyl Acetate and Methadone as Maintenance Treatments for Heroin Addicts," *Archives of General Psychiatry* 33 (1976):709–720.

35. J. Markham, "Nixon Says Heroin Public Enemy No. 1," *The New York Times,* 27 October 1969.

36. Zinberg, "Crisis in Methadone Maintenance."

37. Aldous Huxley, *Brave New World* (New York: Harper & Row, 1969).

38. John Stuart Mill, *On Liberty,* (London: John W. Parker & Son, 1859), pp. 22–23. One may speculate from the larger context of Mill's writing—especially in his discussion of the sovereignty of the individual—that he did not anticipate the extent to which environmental hazards, inter alia, would assault the individual. Otherwise he might not have written: "Over himself, over his own body and mind, the individual is sovereign." Surely we know this, in a scientific sense, to be palpably false. Even the body of the most reclusive hermit, in the event of an explosion in a nearby nuclear reactor, will be affected by that accident.

39. Milton Terris, "Epidemiology of Cirrhosis of the Liver: National Mortality Data," *American Journal of Public Health* 57 (1967):2076–2088. Reginald G. Smart, "The Effect of Licensing Restrictions during 1914–1918 of Drunkenness and Liver Cirrhosis Deaths in Britain," *British Journal of Addictions* 69 (1974):109–121; and Jan DeLint and Wolfgang Schmidt, "Consumption Averages and Alcoholism Prevalence: A Brief Review of Epidemiological Investigations," *British Journal of Addictions* 66 (1971):97–107.

40. Thomas S. Szasz, "The Ethics of Addiction," *Harper's Magazine,* April 1972, pp. 74–79. Reprinted with permission of the author.

41. M.M. Glatt, "Heroin Maintenance?" *International Pharmacopsychiatry* 7 (1972):61–63.

42. Forest S. Tennant, Jr.; Kenneth D. Krantz; and Emanuela I. Dobrin, "Use of Difenoximide (SC-26100) for Narcotic Detoxification: A Preliminary Tolerance and Efficacy Study," *American Journal of Drug and Alcohol Abuse* 4 (1977):123–135.

43. Forest S. Tennant, Jr.; Barbara A. Russell; Judith Tate; and Rich-

ard Bleich, ''Comparative Evaluation of Propoxyphene Napsylate (Darvon-N) and Placebo in Heroin Detoxification,'' *International Journal of Addictions* 12 (1977):565–574.

44. Forest S. Tennant, Jr. and Joseph A. Shannon, ''Detoxification from Methadone Maintenance: Double-Blind Comparison of Two Methods,'' *Drug and Alcohol Dependence* 3 (1978):85–92.

45. Andrew A. Sorensen, ''Mutual Education of Drug Addicts and Townspeople: Establishing a Residential Facility in a Nonurban Area,'' *Journal of Drug Education* 4 (1974):7–12.

46. Zinberg, ''Crisis in Methadone Maintenance.''

47. Vincent Dole and Marie Nyswander, ''Methadone Maintenance Treatment: A Ten-Year Perspective,'' *Journal of the American Medical Association* 235 (1976):2117–2119.

13 Setting Environmental Standards for Toxic Chemicals

Leo J. Hetling

Most people are baffled by the complex structure of synthetic organic contaminants. However, these compounds are the basis for much of the standard of living that we experience today. The telephone, airplane, computer, food-preservation system, and many other aspects of modern life could not exist without synthetic organic chemicals.

The manufacture and use of these compounds in New York State is widespread. During their manufacture and use, some fractions of these compounds do find their way into the environment and into water-supply systems. The chemicals have been found as contaminants in sources of water used for drinking throughout New York State. The presence of these chemicals in water represents some risk to the consumer. One should note that this is not a voluntary risk such as driving a car or smoking a cigarette but an involuntary risk like breathing air or eating food.

In most cases, no acceptable scientific data are available relating low doses of these compounds to observed clinical effects in humans. Because of this, an estimate of risk can only be obtained via extrapolation from animal data or effects of similar compounds. Table 13–1 outlines the types of data used and some of the problems with their use. It is important to understand that these extrapolation procedures remain an area of great scientific controversy. The scientific community not only does not agree on what extrapolation procedure is proper, but also many scientists question the validity of any extrapolation. When extrapolated by the methods adopted by the New York State Department of Health, the magnitude of these risks appears comparable to many other voluntary and involuntary risks (table 13–2).

A variety of engineering methods can be used to control synthetic organic chemicals in drinking water, including the development of new, uncontaminated supplies or treatment by aeration and granular-activated carbon. The cost and efficiency of these measures vary widely but they can be quite high (see table 13–3). For comparison, a recent survey indicated that the present per household cost of water in New York State was less than $150 annually.[1]

A variety of regulatory procedures are available that New York State

Table 13–1
Data Used to Calculate Risk

Basis of Extrapolation	*Problem*
Epidemiological studies	Studies do not exist for most compounds of interest. Data are usually from high-dose studies that must be extrapolated to the low levels actually found in the environment. Studies are usually based on occupational exposure of a given type of individual (white, working-age, healthy male). This must be extrapolated to a normal population.
Animal studies	Low-dose long-term studies are impractical, thus data available are from high-dose short-term experiments. This must be extrapolated to effects of low dose. This requires acceptance of the premise that human response will be similar to animals studied.
Chemical similarity	This depends on the unproved hypothesis that chemically similar compounds act in a toxicologically similar manner. Data on similar compounds often do not exist.
Class value	The method assumes that all compounds in a similar chemical class have similar toxicological properties.

Note: No numerical method for risk extrapolation exists to account for synergistic effects, genetic damage, or teratogenic effects.

Table 13–2
Risk of Fatality by Various Causes

Accident Type	*Chance per Lifetime*
Hurricanes	2.8×10^{-5}
Lightning	3.5×10^{-5}
Air travel	70×10^{-5}
Drowning	230×10^{-5}
Motor-vehicle travel	$1{,}750 \times 10^{-5}$
PCB (polychlorinated biphenyls) in drinking water (0.16 microgram per liter)	0.1×10^{-5}
Vinyl chloride in drinking water (11 micrograms per liter)	11×10^{-5}

could adopt to provide a basis for controlling the level of synthetic organic chemicals in the state's drinking waters:

Require all public water systems in the state to install the best available treatment technology.

Require the larger public water systems in the state to install the best available treatment technology.

Require treatment to reduce or eliminate contaminants in only those

Table 13–3
Comparative Treatment Costs for 1,000 Gallons of Water

	Dollars per Household per Year (Three People per Household)		
Plant Size[a]	*Groundwater Development*[b]	*Aeration*[c]	*Granular-Activated Carbon*
0.1	33	23	—
1	10	10	110
10	—	3	47
50	—	2	25

[a]In million gallons per day.

[b]Uncontaminated source not always available.

[c]Not an effective method of treatment for many chemicals.

systems whose water presents an unacceptable risk because of organic compounds. This requires a public-policy judgment of an acceptable involuntary-risk level.

Advise the public of the levels in, and the risk associated with, their drinking water, allowing them to decide if the risk is acceptable or unacceptable.

Attempt to control contamination by restricting the use or by strictly regulating the manufacture, use, and disposal of organic chemicals.

Adopt the federal standards and/or regulations when they are promulgated.

The practicality, usefulness, acceptability, and need for any one or combination of these strategies is a question of considerable debate.

Although the policy (set of strategies) that will finally be adopted requires scientific input, the selection of this policy is partly a public decision involving a balancing of acceptable involuntary risk against increased regulation and cost. It may also be particularly an ethical, moral decision since, with its adoption, government is purportedly exposing a population to a calculated risk that may result in premature death to some individuals.

Realizing that public comments and suggestions are required before a final strategy for regulating synthetic organic chemicals in drinking water can be adopted, the state's Department of Health has published a report on the issues and has held public hearings.[2] The results of the hearings and the hearing officer's recommendations have also been published.[3]

Notes

1. Unpublished data, ''Water Supply Needs Survey,'' Bureau of Public Water Supply, New York State Department of Health, Albany, May 1981.

2. Nancy K. Kim and Daniel W. Stone, *Organic Chemicals and Drinking Water* (Albany: New York State Department of Health, January 1980).

3. Langdon Marsh, *Report to the New York State Department of Health, Hearings on Organic Chemicals in Drinking Water* (Albany: New York State Deprtment of Health, January 1981).

14 On Markets for Risk

Mark Sagoff

What are we afraid of? If you ride a bus in Washington, D.C., you might find out. Huge red letters scream to you:

STAND BACK OF WHITE LINE,

DO NOT TALK TO DRIVER WHILE BUS IS IN MOTION,

EMERGENCY EXIT,

KEEP AWAY FROM DOORS.

A Martian, were he or she or it to ride a public bus, might guess that we fear traffic accidents more than anything else. Yet this is plainly false. We may go to great lengths to make buses safe, but when we get into our own cars, we do not even bother with the seat belts. We do very little about drunk driving although, according to government estimates, alcohol is responsible for almost one-fourth of the 50,000 traffic deaths that occur in the United States each year.[1] While virtually ignoring this major source of death and injury, we willingly spend about $5 million per life saved in protecting ourselves from the hazards of radiation.[2] And we are not alone in swallowing the camel of risk while straining at the gnat. The British, for example, spend 2,500 times more money per life saved in the pharmaceutical industry than in agriculture.[3]

The anomalies toward risk that arise in public policy are repeated over and over again in private decisions. People who engage in dangerous sports—parachuting and mountain climbing, for example, may refuse to touch coffee for fear it contains a carcinogen. Likewise, a friend of mine who chain smokes makes sure there are no chemicals in her food. I suppose this attitude is made large in public policy because, while the law permits cigarettes to be advertised in newspapers and magazines, it prohibits any additive in food that can be linked with cancer.[4]

The greatest anomaly in our attitude toward risk may be historical. People are exposed to far fewer hazards today than they were fifty or a hundred years ago. Yet we seem to worry about the dangers in our environment now more than ever. Diseases that were scourges at the turn of the century—pneumonia, influenza, tuberculosis—are comparatively innocuous today. I need not mention what food and medicine were like before the regulation of food, drugs, and cosmetics early in this century.[5] Yet we have become more, not less, suspicious. As one observer says:

> One day we hear about the danger of mercury, and run to throw out cans of tuna fish from our shelves; the next day the food to shun may be butter, which our grandparents considered the acme of wholesomeness; then we have to scrub the lead paint off our walls. Today, the danger lurks in the phosphates in our favorite detergent; tomorrow the finger points to insecticides, which were hailed a few years ago as saviors of millions from hunger and disease. The threats of death, insanity, and—somehow even more fearsome—cancer lurk in all we eat or touch.[6]

I imagine that more of us are in a doomsday mood, expecting, as it were, the end of the world or something like it, now than a century ago, when millennial movements were strong.[7] "How extraordinary," Aaron Wildavsky, the economist, writes. "The richest, longest-lived, best-protected, most resourceful civilization, with the highest degree of insight into its own technology, is on its way to becoming the most frightened."[8]

Anomalies

Anomalies in the way we deal with risk—if that is what they are—have given rise to a new intellectual discipline called risk-benefit analysis.[9] It concerns itself with the proper management of societal risks, especially in the area of public health and safety. Experts in this field tend to talk in terms of trade-offs.[10] How much is a life worth, or how much are we willing to pay to reduce risk by a certain amount?[11] How much GNP, consumer satisfaction, or welfare are we willing to forgo for how much safety? These are the kinds of questions economists ask.[12] They also typically recommend that risk management be made cost-effective. Public policy should attempt "to equate the marginal costs of saving lives across programs for a certain number of lives saved. . . ."[13]

The programs that make economic common sense, however, are plainly at odds with much of our legislation.[14] Many of our basic laws, including the Occupational Health and Safety Act[15] and the Delaney Amendment,[16] apparently aim at safety even at the expense of overall efficiency. The questions the public asks, moreover, are not always the same ones that concern economists. People ask about the nature of a risk: Is it in our food or water? Is it a hazard workers undergo in order to keep a job? Do we share equally in the risk, or does it fall primarily on the powerless or on the poor? Is the risk natural, as is the chance of an earthquake, or is it technological, like the risk of a meltdown? These are questions we ask; and the answers may matter more to us than anything we are told about relative benefits or costs.

I think it is fair to say that almost all of us voluntarily accept, even seek, all kinds of risks we would never dream of imposing on anyone else. We would fight, perhaps, before such risks were imposed on us. It is one thing to jump voluntarily; it is another to be pushed, even if the risks and rewards are the same. Californians who are finicky about what Nestlé has put in their eggnog build their patios on the San Andreas fault. They want the eggnog, like the earthquake, to be natural. I am not sure what this means, but it is not a new sentiment (taboos about pollution are as old as history), and it is important.[17]

The risks we are willing to take as members of the public seem to belong to a different order than the risks we voluntarily accept as private individuals. We insist on stringent fire protection in hotels and hospitals while we do little about protecting our own homes. We require airplanes to be completely safe while we risk our lives constantly by failing to maintain our automobiles. We smoke but are upset to discover minute traces of some possible carcinogen in our drinking water. Many factors enter in. Chauncy Starr, a dozen years ago, identified what is probably the most important one. The public, he wrote, is willing to accept voluntary risks that are a thousand times greater than involuntary ones.[18]

Is the Public Irrational?

Is the public, then, irrational? The following argument suggests that we are.

1. Assume that what we want to achieve is the most safety at the least cost.
2. Assume as true by definition that people are irrational (all things being equal) if they reject an efficient means to reach their goal and prefer a much less efficient or even totally ineffective one instead.
3. The efficient way to achieve our goal is to minimize the marginal cost of each unit of safety added by equalizing the cost per life saved or accident avoided over all programs. (In other words, spend money in ways that will produce the most overall safety.)
4. Consider two risks, one voluntary and the other involuntary. Suppose that they pose the same danger—for example, one is benzene at home, while the other is benzene in the workplace. Suppose, moreover, that it is a thousand times more expensive to lower exposure to benzene in the workplace (where the chemical is a necessary element in industrial processes) than in the home (where substitutes can easily be found).

From premises 3 and 4, we may conclude

5. It is very much more efficient (by a ratio of 1,000 to 1) to lessen the risk at home than in the workplace.
6. "The indications are that the public is willing to accept 'voluntary' risks roughly 1,000 times greater than 'involuntary' risks."[19] The public, then, would permit the voluntary risk at home while preventing the involuntary risk in the workplace.

With premises 2 and 5, we then conclude that the public is irrational.

Some experts have reached this conclusion—namely, that in matters of risk-management, the public is irrational. These experts have proposed, therefore, that they should make public policy based on the best available scientific studies and the finest techniques of cost-benefit analysis.[20] Industries, in this spirit, have petitioned the courts to overturn so-called irrational legislation on cost-benefit grounds.[21] (Litigation over the OSHA benzene[22] and cotton dust[23] standards are only two of many examples.) President Reagan, by executive order, has instructed all departments and agencies that "[r]egulatory action shall not be undertaken unless the potential benefits to society from the regulation outweigh the potential costs to society."[24] Some of the president's advisers have said that the courts should invalidate all laws that are not cost-beneficial—that is, that are inconsistent with overall economic efficiency.[25] There are as many ways to force people to be rational as there are ways to force them to be free.

If we accept the first premise of the argument I presented, then I believe we should also accept the conclusion; but is this premise true? I want to suggest that it is not. If the Toxic Substance Control Act,[26] and the Clean Air Act,[27] and the Clean Water Act[28] were intended principally to achieve economic efficiency—to obtain health and safety only insofar as benefits outweigh costs—then much that is in these laws appears to make no sense.[29] Environmental law, however, need not be judged in relation to any economic goal, whether it is health, wealth, welfare, safety, or conservation. It may sometimes be better understood as an attempt by people simply to control the circumstances or conditions under which they lead their lives.

Freedom or Autonomy?

A distinction between freedom and autonomy will help make this point. You are free, according to a traditional conception of freedom, insofar as you can do or get what you want. This popular, if rough or crude, conception of freedom needs qualification, of course, but it will do for our purposes.[30] You may see a connection between social efficiency and this conception of freedom. An efficient society—a society that maximizes the satisfaction of preferences—helps individuals get what they want. Thus efficiency enhances one kind of freedom. This is not the political freedom we have to make laws for ourselves—laws that may express our ideals or most considered values. It is a freedom to satisfy wants. An efficient society need not be a happy one; no one has shown that the satisfaction of preferences leads to happiness rather than to disappointment.[31] If freedom has to do with the satisfaction of preference, however, then an efficient society, since it satisfies desires, promotes freedom as well.

Freedom in this context, however, differs from autonomy. If freedom consists, roughly, in getting what you want or doing as you like, then autonomy, roughly speaking, consists in your ability to get or to do these things on your own, without being beholden to any other person, without accepting favors, and without having the important decisions made by somebody else. If you have to depend for your welfare on the wisdom, expertise, or good will of others, if they determine what you want or how you are to get it, then you lack autonomy, even if your benefactors are wiser than you and give you the best deal they can. If you rely instead upon your own actions and decisions—if you pursue your goals under rules that are neutral and apply equally to all—then you may have autonomy even if you fail to achieve your ends. Duncan Kennedy writes that a person is free to the extent that he satisfies his preferences. Freedom in that sense is coincident with welfare. "Autonomy, by contrast, does *not* refer to the degree of satisfaction, but to the *character of the circumstances on which satisfaction is dependent.*"[32]

Statements concerning an environmental ethic, properly understood, may have more to do with the autonomy of individuals than with the efficiency of society. "A thing is right," Aldo Leopold wrote, "when it tends to preserve the integrity, stability, and beauty of the biotic community. It is wrong when it tends otherwise."[33] The appeal of this statement, I believe, depends on the belief that we all compete as equals under rules that nature sets: Gravity and the carbon-dioxide cycle apply to everyone equally; therefore, these violate no one's autonomy. An artificial environment introduces human control and, with it, the distinction between those who manipulate and those who are manipulated. We may have more welfare in a more artificial or a more controlled than in a natural environment, but we are not likely to have more autonomy. The conditions of our well-being, even if they are enhanced in a highly technological rather than a simpler society, tend to be less in our own hands.

E.F. Schumacher, in his book *Small Is Beautiful,* hit upon a formula that appeals to a vast constituency.[34] We need not argue whether small is efficient—whether, for example, solar and wind power may generate more electricity more safely and more cheaply than nuclear reactors. That question, I believe, is somewhat beside the point. Small is not beautiful, necessarily, because it is efficient; it is beautiful because it encourages autonomy. It gives people control of the means by which they satisfy their desires, and this, I believe, may be more important to them than whether the desires are satisfied. (Does this thought lie behind the Marxist idea of giving workers control of the means of production, which could include worker control of nuclear power plants?) The nicest thing about solar panels and windmills, from the environmentalist perspective, is that you or I can decide whether to install them. The bad thing about nuclear power is that the decision is somehow up to "them."

I want to suggest that people in the environmental, antinuclear, and consumer movements are less concerned about freedom than autonomy. They want to determine what risks they take; they do not want these risks to be determined by others. It does not matter how cost-beneficial those risks are; it is a question, rather, of who controls them. To environmentalists of this persuasion, there is only one sort of acceptable risk. It is not necessarily a risk, the benefits of which exceed the costs; it is a risk that people understand and to which they do, in fact, consent.

Efficiency and Consent

Those who advocate market approaches to public-policy problems do so for at least two reasons. They say that markets, if free and fair, are unanimous consent arrangements.[35] This is a tautology, indeed, if any transaction that affects individuals who do not consent to it is counted as a market failure. They also say that any transaction to which two rational self-interested parties consent benefits both.[36] This, too, appears to be a tautology, given plausible assumptions about the nature of rational self-interest. I do not want to quarrel here with these two conceptual arguments in favor of free and fair markets; I only want to point out that they are different arguments. One justifies markets in terms of consent; the other, in terms of aggregate efficiency or satisfied consumer demand.

The two arguments together give us an important principle—namely, that unanimous consent arrangements such as characterize free and fair markets lead to efficient or optimal outcomes. This is the conceptual basis of arguments Milton Friedman and others have offered in favor of capitalism.[37] Since these arguments rest, as I have suggested, on a system of tautologies or definitions, I have no philosophical fault to find with them. The way to criticize them would be to see whether they apply or ever could apply to anything that exists or is likely to exist in the United States or elsewhere. The problem that concerns me, however, has to do with centralized, or public, decisions, like the distribution of risk associated with energy development and waste disposal, that people do not consent to through market transactions. What, then, is the role of market or cost-benefit analysis? When markets fail, or when relevant markets exist only in the imaginations, surveys, and writings of economists, what is the relationship between efficiency and consent?

I want to suggest that there is none. Economists argue, as I have said, that consent leads to efficiency; the argument is a conceptual one. On the basis of this argument, however, economists tend to assume, perhaps too hastily, the converse—namely, that efficiency leads to, or can stand as, a suitable substitute for consent.[38] Economists who apply market-based analyses in areas in which markets do not exist, or where market failure is endemic, tend to assume that what a perfect market would do in those areas is what people would consent to. However, this rests on an assumed, or an enthymatic, premise that is false. The premise is that a market should exist in the problematic area and that people would consent to making transactions within such a market.

Markets and Risk

Americans, as I have mentioned, swallow that camel of risk while straining at the gnat. We casually accept enormous risks, for example, while driving, that we might avoid at a trivial cost. Yet we insist upon making great sacrifices to prohibit public risks that are comparatively small. We pollute our homes freely with solvents, sprays, cleaners, insecticides, microwaves, and leaking gas. The indoor environment, as a result, is much less safe than the outdoor environment. Yet we cry out against comparatively small amounts of pollution in the nation's lakes, air, and streams. Why? Why do we freely accept voluntary risks and shun those that are imposed on us? Why do we seem to fear some dangers so much more than others?

Economists who believe that much of our behavior under risk is irrational may take another view of it were they to consider that it may be motivated not by fear but by resentment. The problem, then, would be to explain not why we fear some dangers more than others but why we resent some dangers more than others. One answer is plain. We resent risks imposed upon us, as members of the public, by those who seek, in doing so, to achieve economic ends. We resent these risks because, in being subjected to them, we are treated not as persons—that is, as ends in ourselves—but more as a mere means for the production of some economic good such as overall efficiency or utility. This is usually not a goal many of us would subscribe to for society or for ourselves; it seems to be imposed upon us by those who take an economic view of regulatory policy. Accordingly, people who resent being subjected to technological risks—for example, dangers associated with nuclear technology—become even angrier when they are told that the risks to the public are small in relation to the benefits. They resent these risks even though they acknowledge that they face greater dangers quite casually, for example, when they drive their cars.

To understand the source of this resentment or, perhaps, a possible basis for it, consider the way many economists establish a value for life. In a typical research experiment, an economist offers subjects a hypothetical sum, say, $1 million, in exchange for running a million-to-one risk of instant death. Almost everyone accepts that bargain. The odds of death may then be increased, the payment decreased, or both, until some indifference point is reached. Economists frequently advocate the use of this trade-off point in establishing safety standards for nuclear reactors and other dangerous technologies.[39]

The problem with this approach may become apparent if we change the offer to involve a risk not of death but of being tortured, sodomized, having one's hand cut off, becoming a slave, or undergoing a needless lobotomy. Here the economist may ask, for example, if you would accept $50,000 in exchange for a 1-to-1,000 probability that you will be enslaved the rest of your life. Suppose you take the hypothetical bargain. Suppose, moreover, a trade-off point is established at which people usually accept a cash payment in exchange for a certain risk of being made prostitutes or slaves. What have we learned about regulatory policy? How can this information possibly tell us anything we could use or want to know?

That experiments of this kind are perverse becomes obvious when we reflect that we would outlaw all such bargains were they to be made, not hypothetically but in fact. If you accept a 1,000-to-1 gamble on being killed for a sum of money and then have bad luck (for example, because the wheel stops at your number), whoever made the offer, nevertheless, would be arrested were he to kill you. It is still murder even if it is efficient. One simply cannot legally set up a market of this kind. Likewise, you cannot raffle yourself into slavery. Again, outside of the dunking booths at fund raisers, where you buy shots at soaking your boss, our society does not allow markets in torture. If actual markets are forbidden in death, slavery, torture, and so forth, what could we possibly learn from hypothetical ones? Why would we want to determine when death, torture, slavery, or whatever is efficient? Why should a civilized society wish to know that?[40]

Individual versus Community Choice

An answer might be that we need to know how much to spend at the margin for life-saving equipment and for safety in the environment and in the workplace. Let us suppose this is so. Why would individual willingness to pay in hypothetical risk markets give us a basis on which to determine this amount? Three reasons come to mind. Individual willingness to pay, one may argue, indicates an efficient trade-off point between safety and cost. This trade-off, second, may be said to be one to which we usually, or on average, consent. Third, one may argue that the number we get (say, somewhere about $300,000 per life saved) seems like a good one and that there is no better way to derive it.[41] All of these reasons, however, are plainly inadequate.

As for the first, we must ask why efficiency should determine the level of risk to be imposed upon the public. Surely we do not torture criminals for confessions when it is efficient to do so. (We do not experiment with surrogate markets to determine what chance of torture people would accept for money, nor would we use such a figure to justify torture.) Likewise, we do not shanghai luckless citizens as slaves on the grounds that the social savings is more than people would pay to avoid such a risk. If efficiency should not be a criterion for the distribution of torture or slavery, why, then, should it be a criterion for the distribution of death? I simply see no answer to that question.

The second reason is no better. The fact that some people give a hypothetical answer to a hypothetical question is not a ground to believe that other people—or even these people themselves—consent, therefore, to an actual risk imposed upon them. Real consent is given in real markets; and people must consent to the market as well as to the trades they make in it. As far as I know, no one at Three Mile Island or at Love Canal or anywhere else outside of economics departments has consented to the idea or to the ideology of a market approach to public safety and health.

Let us suppose, finally, that we need a trade-off number, although this is itself a questionable assumption.[42] Is individual willingness to pay the best or, indeed, the only way to establish it? Plainly, not. The use of the surrogate-market approach has the merit of setting up juries of citizens; but the question it asks them is the wrong one. They should be asked not to report their self-interest, but, after weighing arguments and listening to reasons, to state their view of the public interest. That information might be useful to the policymaker. How much does the individual think the community should be willing to pay for ambulances? How much for anti-smoking programs? What do we think the responsibility of industry should be with respect to the safe disposal of its chemical wastes? How much should the corporations be willing to invest of their profits for reducing airborne hazards, or for keeping groundwater uncontaminated? When juries of citizens make decisions in civil and criminal cases, they act not on their rational self-interest but on reasons of another sort, namely, the law, the evidence, and a conception of right and wrong. Public officials should be guided by similar considerations. Their job is not to maximize wealth but to respond to many values, among which efficiency may be one, in making and applying public law.

To ask how much people are willing to pay or to take for avoiding or accepting a risk is to change their status from that of members of the public, concerned with the nature of the society they live in, to that of parties in a private contract, concerned only with how a transaction affects them. Individuals may resent the a priori switch of their values and decisions from the public to the private realm; they may not consent to the market they are compelled to make decisions in. To force people to think about public policy in market or in risk-benefit terms is to compel them to accept a particular conception of freedom—a conception inappropriate to the public decisions to be made and inimical to the political institutions in which we should make them. These institutions, in principle, give us freedom to choose not so much what to buy as individuals but what to be as a society—a freedom to associate for common purposes and to decide the nature, market, or nonmarket of our relationships with one another.

I have argued, in effect, that a vast difference exists between choice based on self-interest and choice based on a conception of the common interest. The relation between the two is not that of the part and the whole; one is not a function of the other. They represent, rather, two logically distinct strategies of choice. The individual who chooses on the basis of self-interest need be given information only about the way a particular policy affects him; he need have no concern about how others are affected or about the nature of the policy itself. The individual who chooses on the basis of a conception of the common interest, on the contrary, is concerned about the total meaning and effect of a policy for the society as a whole. A market approach to public policy, as one commentator has noted, "is so set up that 'one's own share' becomes the be-all and end-all."[43] But the second approach, in which the individual is asked to debate and then to vote for a conception of the common will, "becomes a vote not just on *shares* but on *policies,* and not just on a distribution as such, but on the principles involved."[44]

Risk or Resentment

The major statutes enabling administrative agencies to regulate environmental and workplace hazards do not in general allow a cost-benefit or market approach to the management of risk.[45] These statutes, on the contrary, carry into public environmental law the practice in private or common law of providing in nuisance cases not damages but injunctive relief. The Clean Air Act, the Clean Water Act, and the Noise Control Act create public entitlements to reasonably clean, safe, and quiet surroundings. Likewise, the Occupational Safety and Health Act entitles workers to a workplace in which they are treated as persons and not merely as factors in the production of wealth. This is what we approve as a society and this is as it should be on moral grounds. It is more important that we respect each other's safety, privacy, and dignity as persons than that we maximize wealth. What is important to us? Shall we treat each other as persons with rights to protection from hazards others create? Shall we regard ourselves and each other only as producers and consumers instead?

I am aware that the legislation we have is sometimes vague. The requirement that the workplace be free of risk insofar as is feasible is a celebrated example of this vagueness.[46] A cost-benefit approach to risk, however, is no more definite than an entitlement approach: the concept of what is efficient is itself vague. This is illustrated by the ease with which economists prepare cost-benefit analyses to support opposing sides in a policy dispute. Second, any legislation that does not itself, *per impossibile,* set the standards for hazards or pollutants, must stipulate the goals to be achieved in qualitative terms. The agencies must then interpret these goals quantitatively. This inevitably allows a good deal of discretion in determining, for example, whether "as safe as feasible" indicates a ten- or a one-part-per-million standard for permissible levels of benzene. The statute, no matter how well written, is unlikely to rule out every standard but one. Thus political and economic factors are bound to operate at the agency level. The point, then, is to make the process at that level as open, fair, and intelligent as possible. This is the reverse of what is achieved when the administrative process is replaced by a cost-benefit analysis reviewed, behind closed doors, at the Office of Management and Budget.[47]

Third, the legislation we have, taken as a whole, is less vague than it might seem to one who looks at the statutes individually. The law as a whole clearly expresses a national policy toward environmental and workplace nuisances and hazards. The goal—the controlling ideal—is that people have a right to be protected from hazards imposed on them without their consent. No right is absolute, of course; economic and other considerations must be taken into account. This does not suggest, however, that the entitlements in question are themselves economic factors to be balanced with other interests on the basis of willingness to pay.[48] It means, rather, that we will show our good faith by going the extra economic mile to protect those rights. I am aware that no one knows exactly how long the extra mile should be; I am aware that the law is unavoidably vague on this point. But it may appear more precise if we recognize that the purpose of much of our law is not simply to protect health and safety as much as it is to vindicate and support our right to that protection. As a result, the extra mile is longer, not necessarily where more lives can be saved, but where the principles involved are more important. The right to be protected from hazards in the workplace, for example, may be more important than a right we may have to be protected from hazards we create for ourselves at home or on the highway. The law, then, correctly requires us to spend more to control certain risks, for example, those created by radiation or in the workplace, where one person may be used as a means to someone else's ends, than other risks, over which the individual exercises some responsibility.

What are we afraid of? Are we afraid that our health and safety are severely threatened by risks associated, for example, with nuclear installations and industrial wastes? Perhaps. I have argued, however, that we may have more reason to resent than to fear these hazards. What we should fear, I believe, is that entitlements created by public law will be regarded as private wants and that rights supported by the most commonplace conceptions of morality will be treated as benefits. The demand for risk-benefit analysis, as a Congressional staff scientist has said, "is the invention of those who do not wish to regulate, or to be regulated. . . . Its primary use in government decision making is to avoid taking action that is necessary or desirable in order to truly protect the health of the public or the integrity of the environment."[49] At stake in risk-benefit analysis and in environmental policy generally is the difference between what we are willing to pay for as individual consumers and what we decide as citizens we owe to each other and our surroundings. The greatest threat from chemical pollution may be to our sense of being a community acting together. Economic approaches to pollution, more than anything, make us fear for that conception of ourselves.

Notes

1. For a table of fatality estimates from various accidental and environmental causes, see P. Slovic; B. Fishcoff; and S. Lichtenstein, "Rating and Risks," *Environment* 21 (April 1979):19.

2. See R.A. Howard; J.E. Matheson; and D.L. Owen, "The Value of Life and Nuclear Design," in *Probabilistic Analysis of Nuclear Reactor Safety,* edited by D. Okrent and E. Cramer, pp. IV 2-1–IV 2-9 (LaGrange Park, Ill.: American Nuclear Society Publication, 1978).

3. C. Sinclair; P. Marstrand; and P. Newick, "Innovation and Human Risk," Center for the Study of Industrial Innovation, London, 1972.

4. An amendment to the Food and Drug Act introduced in 1978 by Congressman James J. Delaney requires that "no [food] additive shall be deemed to be safe if it is found . . . after tests . . . to induce cancer in man or animal." [21 U.S. Code of Federal Regulations 409(c)(3)(A)]. The principle is extended to animal feeds and food coloring in sections 512 (d)(1)(H) and 706(b)(5)(B).

5. Federal Food, Drug, and Cosmetic Act, 21 U.S.C. §301 et seq.

6. E. Rabinowitch, "Living Dangerously in the Age of Science," *Bulletin of the Atomic Scientist* 28 (1972):5.© 1020-24 E. 58th Street, Chicago, Illinois 60637. Reprinted with permission.

7. See J. Maddox, *The Doomsday Syndrome* (New York: McGraw-Hill, 1972).

8. A. Wildavsky, "No Risk is the Highest Risk of All," *American Scientist* 67 (January/February 1979):32.

9. For an introduction to risk-benefit analysis, see W. Rowe, "Introduction to Risk Assessment," in *Energy-Risk Management,* edited by W. Rowe and G. Goodman, pp. 7–19 (New York: Academic Press, 1979).

10. See, for example, R. Schwing, "Trade-offs," in *Societal Risk Assessment: How Safe is Safe Enough?,* edited by R. Schwing and W. Albers, Jr., pp. 129–142 (New York: Plenum Press, 1980).

11. For discussion of these questions within the framework of risk-benefit analysis, see R. Thaler and S. Rosen, "The Value of Saving a Life: Evidence from the Labor Market," in *Household Production and Consumption,* edited by N. Terleckyj (New York: Columbia University Press, 1976); and J. Linnerooth, "The Evaluation of Life-Saving: A Survey" (Joint IAEA/IIASA Research Report, International Atomic Energy Agency, Vienna, Austria, 1977).

12. For a collection of relevant studies, see Committee on Public Engineering Policy, *Perspective of Benefit-Risk Decision Making,* (Washington, D.C.: National Academy of Engineering, 1972).

13. This statement by A. Kneese is reported in Schwing and Albers, eds., *Societal Risk Assessment,* pp. 144–145.

14. The economic common-sense view of pollution policy may be found in L. Ruff, "The Economic Common Sense of Pollution," *Public Interest,* no. 19 (Spring 1970);69–85.

15. The Occupational Safety and Health Act of 1970 defines an occupational safety and health standard as a standard that is "reasonably necessary and appropriate to provide safe and healthful employment." If toxic materials or chemicals are involved, the secretary of labor is required to "set the standard which most adequately assures, to the extent feasible, on the basis of the best available evidence, that no employee will suffer material impairment of health or functional capacity" [§652(8), 655(b)(5)].

16. See note 4.

17. For discussion, see M. Douglas, *Purity and Danger* (London: Routledge & Kegan Paul, 1970). See also Douglas, "Environmental Risk," in *Implicit Meanings,* Routledge & Kegan Paul, London, 1975.

18. C. Starr, "Social Benefit versus Technological Risk," *Science* 165 (19 September 1969):1232–1237, especially 1237.

19. Ibid., p. 1237.

20. See William Lowrance, *Of Acceptable Risk: Science and the Determination of Safety* (Los Altos: Kaufmann, 1976).

21. Major cases in which industry requested relief from regulation on cost-benefit grounds include Aqua Slide 'n Dive v. Consumer Product Safety Commission, 569 F.2d 831; American Iron and Steel Institute v. OSHA, 575 F.2d 825; Certified Color Manufactures Ass'n. v. Matthews 543 F.2d 284; and Industrial Union Department v. Hodgson, 499 F.2d 467.

22. Marshall v. American Petroleum Institute 100 S. Ct. 2844 (1980).

23. American Textile Manufacturers Institute v. Marshall, Docket No. 79-1429, at present awaiting a decision by the Supreme Court. The current guess is that the Labor Department, now under the Reagan administration, will admit error and withdraw the case.

24. This executive order was circulated to the agencies in late February 1981 but has not as of this writing been published among the presidential documents. Quote is at p. 2 in memograph. (It is to be published as Executive Order 12291).

25. Members of President Reagan's task force on regulation have taken this position. "On regulatory reform, Mr. Weidenbaum of Washington University, chairman of the task force, said he favored an executive order requiring agencies to conduct cost-benefit analyses before adopting regulations. . . . James Miller of the American Enterprise Institution, another task-force member, said he thought all agencies should be required to prove that the benefits of a proposed regulation exceed its cost, and that it has chosen that [sic] least costly alternative. He also said that regulatory analyses should be mandatory and subject to judicial review" [*National Law Review* 3 (17 November 1980):9].

26. Toxic Substance Control Act of 1976, Pub. L. No. 94-469, 90 Stat. 2003, 15 U.S.C. §2601 (1976).

27. Clean Air Act Amendments of 1970, Pub. L. No. 91-604, 84 Stat. 1676 (1970), revised and recodified as the Clean Air Act of 1977, Pub. L. No. 95–95, 91 Stat. 685 (1977).

28. Clean Water Act of 1977, Pub. L. No. 95-217, 91 Stat. 1566 (1977).

29. For example, Title I, Part C, §160-169 of the Clean Air Act as amended (1977) classifies large areas of the West as Class I areas in which "significant deterioration" of air quality is not permitted. This effectively forecloses the development of important energy resources without regard to the benefits or costs. Likewise, much of our environmental legislation is technology based, requiring the use of the "best reasonably available control technology" rather than the technology that will be cost-effective or cost-beneficial.

30. I agree with Gary Watson, who writes, "To circumscribe a person's freedom is to contract the range of things he is able to do. I think

that, suitably qualified, this account is correct'' [''Free Agency,'' *Journal of Philosophy* 72 (24 April 1975):205]. Watson qualifies his account by arguing that a person is free only to the extent that he acts upon motives that are also values or that express his valuational system. Watson characterizes a person's valuational system as ''that set of considerations which, when combined with . . . factual beliefs . . . yields judgments of the form: the thing for me to do, in these circumstances, is *a*'' (p. 215). Watson's thesis is entirely consistent with the view argued by Charles Taylor that responsibility for oneself requires one to be a critic and evaluator of one's own motives [''Responsibility for Self,'' in *The Identities of Persons,* edited by A. Rorty, pp. 281–299 (Berkeley: University of California Press, 1976)].

31. There is an ambiguity in the term *satisfaction* as it is used in economic and moral theory. To say that preferences are satisfied is to say that they are met or fulfilled; the term is used in the same way as when one says that conditions or equations are satisfied. This has no logical connection with satisfaction in the sense of pleasure or happiness. This is the reason that contemporary welfare economics may have no basis in classical utilitarianism—unless one could show that a Pareto-optimal situation is one in which people are happier than they are when they merely strive for but do not have as much of what they want. That the satisfaction of desire does not lead to pleasure but to frustration and disillusion, while fantasizing instead makes one happy, is a commonplace of poets (for example, John Keats, ''Ode to a Grecian Urn,'' and Emily Dickinson, ''Success is Counted Sweetest''). For this idea in economic theory, see W. Weisskopf, *Alienation and Economics,* (New York: Dutton, 1971).

32. D. Kennedy, ''Legal Formality,'' *Journal of Legal Studies* 2 (1973):371.

33. Aldo Leopold, *A Sand County Almanac* (New York: Oxford University Press, 1949), pp. 224–225.

34. E.F. Schumacher, *Small Is Beautiful: Economics as if People Mattered* (New York: Harper & Row, 1973).

35. ''In the first place, relationships in the market are a form of unanimous-consent arrangement'' [Charles Schultze, *The Public Use of Private Interest* (Washington, D.C.: Brookings Institution, 1977), p. 16].

36. ''A most important point that I wish to make about the phenomenon of exchange is that not just one but both parties to an exchange are better off after the exchange is executed than they were before'' [William Baxter, *People or Penguins? The Case for Optimal Pollution* (New York: Columbia, 1974), p. 19].

37. Milton Friedman, *Capitalism and Freedom* (Chicago: University of Chicago Press, 1962), chapter 1.

38. For a presentation of the assumptions behind this move from what

efficient markets would do to what governments ought to enforce, see Allen Kneese, "Environmental Policy," in *The United States in the 1980s,* edited by Peter Duigan and Alvin Bavushka, pp. 253–283 (Stanford, Calif.: Hoover Institution Press, 1980). For a typical discussion of the way Pareto notions of improvement or efficiency connect consent with the outcome of cost-benefit analysis, see Edward Gramlich, *Cost-Benefit Analysis of Government Programs* (Englewood Cliffs, N.J.: Prentice-Hall, 1981), especially chapter 4.

39. Graham Green's *Dr. Fischer of Geneva or the Bomb Party* (New York: Simon & Shuster, 1980) tells of a party in which guests are invited to pull open Christmas crackers that either contain a sizable check or a lethal bomb. The party is the idea not of an economist but a misanthropist in order to cause people to humiliate themselves out of greed and thus reinforce and vindicate his contempt for the human race. For a discussion of degrading or perverse preferences (though not one that concerns economic approaches to risk) see Brian Barry, "Lady Chatterley's Lover and Dr. Fischer's Bomb Party: Liberalism, Pareto Optimality and the Problem of Objectionable Preferences," forthcoming in *Foundations of Social Choice Theory* edited by Jan Elster (Cambridge: Cambridge University Press). See also Robert Goodin, "Laundering Preferences," in the same volume.

40. "Consistency with the criterion of potential Pareto improvement and, therefore, consistency with the principle of evaluation in cost-benefit analyses would require that the loss of a person's life be valued by reference to . . . the minimum sum he is prepared to accept in exchange for its surrender" [E.J. Mishan, *Cost-Benefit Analysis* (New York: Praeger, 1976), p. 305]. This statement occurs in a discussion of ways to assess willingness to pay as a criterion for establishing a social valuation of risk for loss of life and limb.

41. Martin Baily and Robert S. Smith, in *Reducing Risks to Life: Measurement of the Benefits* (Washington, D.C.: American Enterprise Institute, 1980), pp. 52–66, survey various benefit-cost assessments to arrive at a plausible lower bound between $170,000 and $300,000 and a plausible high bound between $715,000 and $3 million. A $287,175 figure is discussed by Fred Hapgood in "Risk-Benefit Analysis: Putting a Price on Life," *Atlantic Monthly* 243 (January 1979):33–38.

42. Environmental and occupational health and safety law in the United States, indeed, is not based on an economic evaluation of life but is technology based, requiring that industries invest in the best (reasonably) available or achievable technology for risk reduction. What is reasonably or feasibly achievable has to do with the state of the art and the ability of industry to survive.

43. Nicholas Rescher, "Economics vs. Moral Philosophy," *Theory and Decision* 10, no. 169 (1979):171.

44. *Id.*

45. For discussion, see M. Baram, "Cost-Benefit Analysis: An Inadequate Basis for Health, Safety, and Environmental Decisionmaking," *Ecol. Law Quarterly* 8, no. 473 (1980). See also M. Sagoff, "Economic Theory and Environmental Law," *Michigan Law Review* 79, no. 1393 (1981).

46. See n. 15 *supra*.

47. See Executive Order Number 12,291, 46 Fed. Reg. 13,1393 (February 19, 1981).

48. For the now-standard analysis of rights as trumps against collective economic interests, see R. Dworkin, *Taking Rights Seriously* (1977, 1978), esp. ch 4.

49. Burke Zimmerman, "Risk-Benefit Analysis: The Cop-out of Government Regulation,"*Trial Magazine,* February 1978, p. 44. Zimmerman was staff scientist for the House Subcommittee on Health and the Environment when he wrote this article.

Part III
Interventions at the End of Life: Choosing, Accepting, and Allowing Death

Almost as difficult as the determination of when life begins is the determination of when life ends. A government commission recently suggested a model for states to consider when writing laws to define when death has occurred. It seems that both ends of the spectrum, life and death, are not as sharply delineated as it appeared and that either government or the law is being asked more and more frequently to produce a clear meaning for these two concepts. It may be asking too much of either institution, or perhaps asking the wrong institution, to attempt this task. In the case of legal institutions, individual patients have already died from natural processes before final appeals result in a ruling on whether to pull the plug or to continue the treatment. A body on life-support systems can be legally dead (according to brain-death statutes) in an ambulance traveling in one state and be resurrected when the vehicle crosses a border into a state that has different statutes. In some states (for example, New York), one reading of the law would necessitate a court order for any but the most heroic of treatments. In the case of legislative bodies or governmental agencies, philosophical and theological issues are to be made state regulations, accepted by all citizens. This amounts to a state religion, of course, one not based on human faith or conviction but on political power and legal leverage. Somewhere in the ethical discussion, both personal and religious faith, and the medical institution with all its values and emphasis on the realities of life and death, have gotten lost in legal and civil rule setting. For these reasons, we think it safe to say that the field of bioethics has not been a constructive influence to the large extent that it has encouraged the use of legal and governmental institutions in areas of attitudes and understandings of life and death.

The area of death and dying, as a result, has been politicized and legalized and the existential and medical experience lost from view. Chapter 18, on the hospice movement, demonstrates that it is the human process of dying, and individual coping, along with the decision that meaning is no longer present (that personhood and physiology have parted company), that is important. It is not a scholastic exercise in when death has arrived. As Paul Hardy (chapter 19) and Richard McCormick (chapter 20) illustrate, our legal institutions are poorly constructed to handle those questions. To ask these institutions about the meaning of existence, the value of living, or the best way of dying is to ask them to solve the purpose of existence for us. Finally, there is even room for a personal decision that existence has no meaning and life no value. In all these areas, bioethics has sometimes mistakenly looked for rules and for a checklist of rationality.

The Ethics of Suicide

Opting for death seems a terrifying choice. Again, the ethical analysis of this choice has focused on whether dying appeared to be the rational thing to do; so-called rational suicide has become descriptive of the issue. It may very well be that the two terms—*rational* and *suicide*—cannot be put together. This depends on whether suicide is viewed:

> As another weighted decision among many a person makes (suicide as decision making),
>
> As a psychotic episode (suicide as psychosis),
>
> As a sick wish to hurt self and others (suicide as aggression),
>
> As a sacrifice for a cause or for another person (suicide as altruistic),
>
> As the expression of a feeling of meaninglessness and valuelessness (suicide as nay-saying).

Rational decision making is ultimately a balance sheet of values. To be rational it must exclude psychotic and highly neurotic definitions of suicide. Altruistic suicide is usually redefined as something other than suicide. Thus, we seem to have a choice, if we wish a general definition of rational suicide, of combining the first and last definitions: rational value choices and ultimate rejection of value and meaning. This is an impossible combination. Thus, there are either two radically different meanings for nonpsychotic and nonneurotic suicide or one or the other description is inaccurate. Clements, Sider, and Perlmutter (chapter 16) will argue that the latter is true and thus that rational suicide is a meaningless concept. Thaler (chapter 15) will maintain the psychotic and neurotic definition holds so that rational suicide makes no psychiatric sense.

If trying to decide what is meant by a clear act of suicide is difficult, trying to determine when a death *is* suicide further complicates the problem. We need to distinguish suicide, if we can, from dangerous risk taking, accidents, and slow self-destruction (alcoholism, smoking, medical noncompliance—that is, destructive life-styles in general).

These complications are all prelude to a determination concerning intervention, which is a different ethical question from the issue of labeling suicidal acts or the person who commits suicide as good or evil. Should we try to prevent suicides, and if so, how?

Suicide has been described as the ultimate ethical question, and certainly Hamlet's question was "to be or not to be?" The Stoics' "open door", and tradition even into the early Christian period, considered suicide an ethically acceptable alternative to difficult life situations. Our own period has seen a more openly hostile element in it. Jim Jones and his group at Jonestown used it to make an angry, aggressive statement about their larger society. There is an asocial character to suicide, as Aristotle understood, that is sometimes combined with a disregard for the effects on close friends or family. Yet being asocial, lacking empathy, negating compassion, and rejecting existence are quite alien to all that makes ethics or values possible. Since this is the point of view of the person wishing to kill himself, it makes little sense to try to talk to him about good and evil, or even beauty or truth, and much less the medical values of well-functioning and survival. This shifts the ethical perspective regarding suicide to the interveners and the social networks and away from the act or the individual.

Otto Thaler focuses on the clinician's perspective in chapter 15, and most important in terms of the discussion in the introduction to this book, on the ethical aspects inherent in the clinical process itself ("Any interaction between persons, any human activity, has a moral aspect"). This is the outcome of clinical necessity, one that grows out of the human interaction that is the physician-patient relationship and that is based on the patient's genuine needs. Thaler defines a clinician, and in so doing defines the expert role as someone who has intimate familiarity with things, systematic knowledge, an effective theory, involvement, and a capacity to act. This leads to responsibility, something clinicians accept almost as second nature, but that others may either overlook or pay lip service to without comprehending. As a psychoanalyst, Thaler views all human conduct as determined (meaningful), but a large part of that determination is unconscious or not readily discernible. Therefore, to give a suicidal patient a real choice, he needs to be assisted to discover the motivations behind the suicide impulse. Thaler's experience is that such a fuller understanding will cause a reevaluation and a choice for life. This is important because suicide is irreversible, final. For the clinician to allow the act, when suicide is primarily motivated by unconscious forces, is irresponsible.

In addition, suicide is not solitary but directed at others. As an aggressive act, it can give rise to aggressive, retaliatory acts from those at whom it is aimed. Thaler labels the professional's allowing suicide as a "retaliatory act of counteraggression,'' revenge for the hostility. He finds suicide innately vicious and employing the rationale of right to choose (autonomy), for example, is only a disguise for that aggression. Suicide is a muted cry for help, finally, and the responsibility for answering it cannot be professionally avoided.

Chapter 16, by Clements, Sider, and Perlmutter, tries to refocus the ethical analysis. They suggest we are mistaken in trying to assign an ethical value to the act or the person considering it. For theoretical reasons, they argue, it is meaningless to assign a value to suicidal choice; suicide is a negation of all values. It is an expression of an *attitude* that is prior to all values because it is the negation of the ground for all values: existence, well-functioning, and pleasure. Not only does this apply to ethical values (the good) but also to cognitive values (truth) and aesthetic values (beauty) as well. For this reason, rational suicide makes no sense, the right to suicide is academic, and whether suicide is good or evil begs the question. The ethics of suicide needs to be focused on intervention, not the act; on the intervener, not the actor; and the justification for intervention will be in terms of the values of the intervener, which must include real compassion and concern for the needs of the would-be suicide.

Since suicide calls into question the medical values of the profession, and in fact, the justification of those values, and questions the working alliance of physician and patient, it is an extremely disquieting problem. Clements and her colleagues argue that to ethically justify intervention requires more than a rational calculus. ''A behavioral and dynamic understanding of each suicidal person we contact, and of our own motives and unconscious forces, must inform the ethical choice.'' The ethical reason for intervening must be a reality-oriented relationship and not a countertransference. Finally, there are issues concerning the nature of intervention: its empathic character, its qualitative nature and extent (what kind of intervention and how costly or coercive), and its temporal nature (when and how long). Intervention is chosen, in conclusion, because of values (well-functioning, existence, potential for life) the intervener holds as part of the human community and tradition. Also very important, but overlooked in many ethical systems, is that as empathic, social beings, interveners feel for others and are motivated by that feeling. When the interveners are clinicians, they are motivated by medical values including health, compassion, alleviation of pain, and systematized knowledge. There are even those motivated by an aesthetic appreciation of a well-functioning physiological and psychological process.

The Good Death

The discussion of choosing to die is based on choosing between alternatives of a good death and a bad death, so we need to look behind these slogans. Daniel Maguire in chapter 17, focuses on terminal illness and considers the possibility of a rational choice to hasten death. The hospice movement, discussed by Dorothy Summers in chapter 18, is based on value judgments about the dying process, acceptance and intervention, and choosing the manner of dying. More broadly, Paul Hardy (chapter 19) and Richard McCormick (chapter 20) address the ideas underlying the style of dying, the notions about death, the limits to the choices we can make about death, and the concerns about choosing for others when it becomes a question of prolongation or quality of life. Again, these issues have been generated by advancing technology, resource limitations, and the involvement of legal and governmental institutions. These have produced an attention to what can be done, how economically costly it will be, and rules for when and how to do it. Perhaps the question that needs to be asked, however, is why and to what purpose we should intervene. That question cannot be answered by law. Cost-benefit analysis only tells us how to implement our purposes, not what they are. Technology makes a wider range of purposes available, broadens our options, but does not do the opting. What the good death is cannot be answered in those terms.

The good death involves choices about prolonging life and the desirability of living. It involves a choice between seeing life as the ultimate value or seeing it as a condition for value. It involves trying to achieve a sense of meaning in the dying process and ways to best insure that. The hospice movement developed out of a sense that hopeless heroic treatment and concentration on technology to the exclusion of human application and interaction was not the best way to achieve meaning in the dying process or to alleviate the most pain and suffering. Human caring and a realistic assessment of what continued treatment can achieve, along with a deep sensitivity to controlling pain, seems essential to a good death. There is a study of terminal burn patients that indicates that when given a choice between full technical treatment and having their hand held, the large majority chose to have their hand held. Thus, the hospice movement is an attempt to insure that this human contact and caring will be present and that the dying will not be isolated and rejected. This is no small task. It is a delicate balance as well, because it is not always easy to predict when therapeutic intervention is no longer productive and when palliative intervention should replace it. It is not always easy to distinguish realistic from unrealistic fear, or purposeful from purposeless discomfort. Patients are also very different and do not go lockstep into that good night, through stages of dying that may not exist as such schemes. Some genuinely need to fight to the last as honest expressions of who they are, some need to deny to the end and have their defenses respected, and others need to control the details of their death. Some are just very tired. A danger exists of searching for the romantic perfect death for everyone. In addition to this danger, Cecily Saunders, director of St. Christopher's Hospice in London, has expressed the fear that the good death will become a duty or obligation of the dying patient and not really a choice. (Note here that such concern about options becoming obligations has also been discussed in the realm of reproductive choices.)

Daniel Maguire (chapter 17) feels the current attention to issues of death and dying is not a morbidity boom but a needed intellectual maturing and acceptance of limit as an ultimate reality. There is wisdom in the ancient ideal of dying well, or the art of dying. Maguire questions the assumption that life in any condition is better than death and uses Karen Ann Quinlan as an example that some things are worse than dying. The Quinlan case could not have occurred without modern technology, which reemphasizes a point made in the previous chapters: our scientific advances have given us new choices, responsibilities, and accountability. The ethical use of our new abilities requires thoughtful action, and sometimes that involves a choice for death. Although life is a precondition for all good, prolongation of life is not always a good (chapter 20 develops this further).

Since it is not physicians, but rather our "cultural uneasiness with death" that is responsible for this problem, Maguire calls for an appropriate-care act, with the suggestion that at some point appropriate care may really be to company with the dying to achieve a good death. From a theological perspective, to say either ceasing heroic treatments or practicing euthanasia is moral or immoral is to present an image of God, and this underlines the seriousness of such ethical analysis. In addition to choices about dying, the hospice movement, gerontology issues, and phenomena of grief and bereavement are part of the picture. Finally, Maguire critiques the "sado-masochistic myth" about suffering, which he says has long obscured Christian understanding of it. The old view saw suffering as having intrinsic value. Maguire offers what he terms a genuine Christian view of suffering as a physical evil, a sign of creation's incompleteness. Thus, scientists who work to alleviate or control suffering are doing God's work, they are cocreators and coredeemers. Value, rather, is in the sufferer's ability to transform suffering and to bring meaning to it. Ending on an optimistic theme, Maguire points out an emerging literature on the afterlife and a shift in consciousness within the culture that, while frightening because new, means a greater understanding of life and death in the future.

Dorothy Summers gives a sensitive, firsthand description of St. Christopher's Hospice in chapter 18. She defines science to include not only heroic technology but also knowledge of how best to care for the dying patient and his or her family. The dying patient is not a failure of medicine; rather, an important part of medicine is the task of caring for that patient to the end. The individual patient is the crucial part of the interaction, and the goal is to control symptoms so that quality of life will make living worthwhile and leave the family with an experience that can be remembered with the positive moments as well as the painful ones. At St. Christopher's, patient status is always reviewed in order to make certain that proper medical management can shift from palliative care back to therapeutic care if indicated. This is essential to avoid the warehousing of dying patients in the interest of cost containment, for example. Pain control is important, with morphine and phenothiazine predominant. The hospice is not only optimistic that pain can be controlled, but it also pays attention to anxiety and fear of death. Dying is not removed from the ward and patients have social support. Bereavement and separation are painful parts of dying, but the hospice helps by breaking down the silence among family members and helping to work toward a closeness even in the process of separating.

However, there are broader concerns than the good death of terminal patients who are assisted in a natural process of dying. For example, issues of hurrying this process, of assisting in it, or in the case of extremely defective newborns, of killing such eventually terminal individuals must also be considered. Questions about pulling the plug, giving no nourishment, refusing a surgical procedure, or killing with an overdose of a drug involve again the choice between life prolongation and quality of life. Issues of who should make these life-and-death decisions if the patient is not competent, and how to determine if the patient is competent, are also crucial.

Paul Hardy (chapter 19) points out that powerful new technologies are two-edged swords, raising new and serious problems and forcing us to reexamine our theological and ethical assumptions about death and dying. With new possibilities for control come new responsibilities. In the area of terminal illness, this requires a new look at legal requirements and ethical proscriptions. Hardy cites a number of legal cases *(Spring, Quinlan, Saikewicz)* to illustrate the difficulties within a legal framework of developing humane guidelines for decisions on continuance or termination of life-prolonging treatments. He develops in detail the *Spring* case, a long and complicated legal proceeding in which legal guardians and proponents of a sanctity-of-life view replaced the wishes of the family and the judgment of the physician. At present, legal precedent runs the spectrum from *Quinlan* to *Saikewicz* (and in New York, we could add *Fox* and *Storrar)*, so the resort to a legal determination of this ethical-medical problem has not been successful, creating confusion over medical duties and responsibilities and the surrogate's (usually, but not always, the family) role. Hardy points out that the court's role was "cumbersome, inefficient, costly, and unnecessarily time consuming."

Hardy turns to a determination of competency as one of the key factors in these cases. Psychiatrists have usually been given the task of assessing mental status. Unfortunately, the legal and psychiatric meanings of mental competence are not identical. Added to the limitations of contemporary forensic psychiatry is what Hardy feels is the limitation of psychoanalysis. He argues that psychoanalytic methods are inadequate for gathering clinical data about mental functioning in organically impaired patients and for organizing data into a coherent framework for evaluation. He suggests behavioral neurology as an alternative. Behavioral neurology seeks to analyze human behavior and intellectual functioning through neuropsychology, and neuropsychology in turn offers the possibility of cerebral localization of mental functioning and uses a battery of objective neuropsychological tests to obtain information. Behavioral neurology, Hardy argues, not only would clarify legal issues of competency but also would have "implications for general theories of ethics and for particular issues in medical ethics with respect to dementing and brain-damaged persons." It has implications for the notion of personhood, expanding it from cognitive capacities to include emotional capacities. Hardy is thus calling for an emotive theory of ethics in place of a narrower cognitive one.

Richard McCormick begins chapter 20 with a discussion of the New York State *Brother Fox* case that established complicated procedures for pulling-the-plug decisions for comatose patients—namely, three doctors must determine that the condition is terminal, the district attorney must be informed, selected doctors must review the case if wished, and then a court hearing is held. This, along with other cases mentioned in chapter 19, forms a new pattern that removes medical management decisions from the patient-family-physician relationship, where they can be personalized, to three impersonal factors: high technology, cost containment, and public entities such as the law and the courts. While courts are sometimes needed, McCormick does not want to start with a presumption against families and physicians, but for them. He then turns to proxy decisions concerning babies maintained on machines. He underlines the need for disciplined convictions about decisions in order to avoid the impersonal forces continuing to produce a "new priesthood of the judiciary." He raises cases of repeated and high-risk surgery on newborns, of meningomyelocele, and of necrotizing enterocolitis and asks when and why concerned parents and physicians would cease treatment. McCormick argues that most bioethical problems involve a protected and promoted basic human value coming into new technological circumstances. For example, childbearing and parenting can now occur under the new circumstance of in vitro fertilization, and the same thing has occurred in the area of preservation of life. The question is: When is preservation a benefit—in the best interest of the patient? The Western tradition is a middle ground between the extremes of medical moral utopianism or vitalism (life at any cost) and medical moral pessimism (when dysfunctional, kill). McCormick states the basic value judgment about this: "Human life is the source of all other experiences and values and is a basic good, the most urgent basic good, but not an absolute one." Traditionally, we do not have to use extraordinary means to maintain life, but he points out that the terms *ordinary* and *extraordinary* have been badly misunderstood. They do not simply mean the difference between what doctors ordinarily do in a crisis, at least not ethically. Rather, they refer to any means used "without excessive inconvenience and that offer reasonable hope of benefit." These are criteria relative to the patients, not absolute criteria. With medical advances, grave inconvenience is becoming less a consideration, so the focus becomes benefit to the patient. In what condition will the patient be left?

McCormick argues that we need to find criteria for determining an actual benefit to the patient. He first discards maintenance of simple body functions (vegetative state), assuming we do not preserve life just for that goal. He suggests two minimum criteria: "(1) absence of severe protracted unmanageable pain and (2) some minimal degree of human interrelating." He adds that in dealing with infants, we need to favor life whenever there is a reasonable doubt about whether these criteria can or cannot be met. With adults, use of documents such as a living will can remove elements of doubt concerning comatose patients. McCormick concludes by seeing a trend to substituted judgments based on assessment by reasonable people and the need for bioethics to assist in making us knowledgeable, reasonable people.

15 The Ethics of Suicide: A Clinician's Point of View

Otto F. Thaler

I have some concern about the title chosen for this book: *Science and Morality*. I bridle at the implied dichotomization of science on the one hand and morality on the other. Any interaction between persons, any human activity, has a moral aspect. A moralist whose thinking does not start with the rational consideration of experiential data ends up a fanatic. A scientist who ignores the moral implications of his work ends up designing more efficient gas chambers.

Sigmund Freud, the most austere moralist of our time, was at the same time totally committed to a rational, scientific world view. Morality is the outcome of necessity, he said. For him, moral issues were, so he told us, for the most part self-evident. The necessity he speaks of is that of the requirements of truly human intercourse. The ethics and the morality of suicide are, to me, the outcome of clinical necessity—a necessity that emerges from the interaction between physician and patient. It is defined by what is required to fulfill the true needs of the patient.

I propose to present a clinician's point of view on the ethics of suicide. Before I do that, however, I need to define for you the term *clinician*. Since I said a clinician, not *the* clinician, I also need to tell you what kind of clinician I am. Hippocrates demands of the clinician:

> Hard, intelligent, unremitting labor in the sickroom or at the bedside. The word *clinician* comes from the Greek *klinē,* a bed.
>
> Accurate observation of things and events and the selection and classification of the data based on observation of the regularity of the events of illness.
>
> Judicious construction of a theory that will allow the ordering and the manipulation of the observed phenomena.

The clinician must have intimate, habitual, intuitive familiarity with things, must have systematic knowledge of things, and must have an effective way of thinking about things. I shall add two additional requirements: involvement and the capacity to act. Involvement here means the capacity for controlled intimacy, the capacity to respond, based not only on the patient's overt demands but also on the patient's clinically inferred, often contradictory, covert, actual needs. This involvement implies responsibility. The capacity to act is what, perhaps, distinguishes the clinician most from more purely academic colleagues than any other professional requirement. The clinician knows that it is not possible to do nothing. Inaction, in the clinical setting, has as profound consequences as action. Inaction can be as deliberate and fateful an intervention as the most dramatic surgical procedure. The clinician is always acting, interacting, influencing, intervening and always on the basis of insufficient, inadequate, incomplete, or sometimes erroneous data.

First, then, I am such a clinician; I am intimately familiar with the raw data of the clinic. I have acquired systematic knowledge of the range and patterns of such data, and I have an effective way of thinking about these data. I have learned how to become optimally involved and have had to accept action and its responsibilities as a basic mode of professional conduct.

Second, I am a psychiatrist—that is, I am a psychiatric clinician. My clinical work is concerned primarily with those dysfunctions of the human organism that are manifested at the psychological level of organization. I consider the human organism as an open biological system of which the psychological level of organization is a completely integrated and simultaneously functioning subsystem. A dualistic, Cartesian, view of mind and body is archaic and clinically useless. It follows that I include dysfunction at the psychological level of organization as potentially part of the disease spectrum. The study and treatment of such dysfunctions, therefore, are part of medicine, the special province of the psychiatric physician.

Third, I am a psychoanalyst. As a psychiatrist, my way of thinking about things, my theoretical framework, is that of psychoanalytic psychology. I have learned that this way of thinking about things is applicable to all patients, indeed to all events at the psychological level of organization. Informed as I am by this way of thinking about clinical phenomena and about myself, I assume that all human conduct is determined—that is, meaningful—though this is not always readily apparent. I know that a large and important sector of the determinants of conduct is unconscious—that is, inaccessible to voluntary recall. I know that among the unconscious determinants are motivations, both of a sexual and aggressive nature—early-life experiences and thoughts, feelings, memories, and fantasies—that, though now out of our awareness, affect our conduct in crucial and fateful ways. The meaning of a psychological event is the intention it serves and its place in a mental sequence—that is, where it fits into the uninterrupted stream of psychological processes. Frequently, more often than not, neither the intention nor the place in the mental sequence is readily discernible to the patient or the physician. As a clinician, my job is to establish both intention and place in the mental sequence—meaning—before I can intervene, or refrain from intervening, in an intelligent fashion.

When confronted with a suicidal patient my questions are: Why? What does this conduct mean? What is its hidden unconscious intention? Puzzling as it may seem, how does it make sense in the context of the patient's life and current state of adaptation, once its unconscious aspects are uncovered? The potential or determined suicide does not know the answers to these questions. Clinical experience has taught me that, once the answers become apparent, the suicidal person, driven till then by poorly understood suicidal impulses, can and most often does reevaluate his condition and chooses life. I consider it my obligation as a clinician to guarantee to the suicidal patient that choice. This may mean, first of all, forcibly preventing the suicidal act.

The patient is about to commit an irreversible and irreparable act. He is ignorant of the true motivation for it and is entitled to the psychological truth about himself. Conduct primarily motivated by unconscious forces is in a sense a lie, a form of dishonesty with self and others—bad faith in Jean Paul Sartre's term. The suicidal act in such persons is, therefore, a desperate act of deception unwittingly perpetrated by the patient. The clinician has highly probable knowledge, not available to the patient, that can alter the patient's perception and hence his plan of action. To allow the patient to proceed without this knowledge is equivalent to allowing a blindfolded man to walk off a cliff because he insists that he knows the road is safe and that he has the right to walk on without hindrance, and with his blindfold in place.

In offering you some examples, I exclude situations such as the decision for suicide in a terminally ill, incurable cancer patient or in politically inspired acts of self-immolation that usually do not come within the purview of the psychiatrist. I have altered certain data so that the patients I tell you about cannot be identified. The situations I describe are common and typical of the kind of problems a psychiatric clinician deals with in daily practice.

The first patient is a 31-year-old single man who is admitted to a psychiatric unit following his attempting suicide by smashing his car into a concrete abutment. He was extricated from the total wreck—unconscious but with only minor physical injuries. On admission he protests vigorously against being detained against his will, expresses regret at having failed in his attempt to kill himself, and gives no guarantee that he will not try it again. Even though he, at first, refuses to engage in any contact with his psychiatrist, he soon relents and can be engaged in a beginning evaluation of his situation.

He is a scientist, employed by one of the local manufacturing concerns, for which he has worked successfully for the past several years. Six months prior to admission he was promoted to a supervisory position, having worked essentially on his own till then. He found no joy in his promotion but accepted it. He soon began to experience difficulties interacting with the people he was supervising, could not tell them what to do, and was increasingly swamped by petty details. He became increasingly despondent and brooded about what he considered his failure as a man and about his loneliness. Life held no further promise for him; he only saw a dreary round of unsatisfactory work and long evenings alone in his apartment. He arrived, so he said, after much deliberation, at the only logical solution: suicide. He was not crazy, he said (true—clinically, he was not psychotic), and he had a right to dispose of his life any way he wanted. He was retained in the hospital against his wishes. What about his family? He came from a lower-middle-class background. His family belonged to a white, ethnic minority. Of several children, he was the only one to have a higher education, which he felt was incomprehensible to his parents; he recalled with bitterness informing his mother that he was finally about to get his doctorate and her obvious disappointment at his getting a Ph.D.—not becoming a "real" doctor. He had had little or no contact with his parents since then. His suicide attempt, it turned out, occurred the day after his birthday. He had received no greetings from anyone in his family. He had called his parents' home the evening before, had let it ring once, and had hung up. He acknowledged later a fantasy of his death's being a feature story in the newspaper and his mother's hearing of it in that fashion.

He was able to develop a relationship with his psychiatrist in the course of three weeks of regular discussions in the hospital. This was accompanied by a gradual diminution of suicidal thoughts. When the relationship seemed solid he was discharged, to be followed on an outpatient basis. More prolonged work produced much insight into intense guilt for outstripping his father and his siblings, and much rage over the lack of support for his endeavors since his early school years. He experienced the people whom he was supervising at work as if they were his younger siblings and was, unconsciously, fearful of what he might do to them. This was one reason for his rapid deterioration after being promoted. Gradually, his depression lifted, his life became a bit less arid, his work situation improved. Follow-up several years later revealed a productive, still somewhat isolated and socially restricted scientist, who seemed content with his life. There was no recurrence of either serious depression or suicidal behavior.

Our second patient is a 21-year-old married woman who, three months after delivering her first child, a girl, took an overdose of sleeping pills. She was found unconscious, was revived, spent a few days in intensive care, and was then admitted to inpatient psychiatry. When interviewed, she was tearful, self deprecatory—"all the grief I am causing my poor husband"—helpless, and hopeless. "All I want is to go home. This place only makes me feel worse, I've got to take care of my baby." At the same time she pleaded: "Why couldn't you let me die, just let me die?" Both requests were denied. That evening she attempted to slash her wrist with the tab from a soft-drink can. She had a deprived childhood; her mother was divorced, working two jobs to raise her and two younger siblings, whose care often fell to her. Patient's husband, also age 21, became less and less attentive to her following the birth of the baby and, over her objections, went out bowling the evening of her suicide attempt.

Further discussions with the patient (who also was not psychotic) revealed her ambivalence about and envy of her own baby: "I am supposed to do for her what no one has ever done for me." There was rage at her "poor" husband for his failure to support her in this difficult time and intense guilt and self-punitive blame for such feelings. Joint meetings with her and the husband began to put some of these issues in perspective; bringing the baby to the hospital afforded some help with mothering skills. She was able to relinquish the suicidal solution to her problem.

Suicide is not a solitary act; it is directed at another. Among its motivational factors are hatred, rage, revenge, envy, jealousy and pleas for attention and love, as well as self-hatred and self-punishment for harboring such feelings. Its intention is frequently the destruction of the image of the other, indeed of the image of the world within the self, the self that is to be destroyed as well. Unconscious as these themes are, the communication of the suicide—I hate you, I want to possess you totally and to destroy you—is clearly understood by those to whom it is directed. These kinds of messages frequently are responded to in kind by neglecting to hear the suicide's cry for help or plea for rescue, or by unconscious collusion with the suicide's purpose. A recent letter to "Dear Abby" reads as follows:

> Dear Abby: Yesterday at work I overheard a young, well-educated co-worker talking to a fellow employee. She said, "My mother is having this middle-age menopause depression. This morning she threatened to kill herself, so I calmly said, 'Mom, in case you've forgotten where the gun is, it's under the pillow on Dad's side of the bed. The gun is loaded, so why don't you quit talking about it and go ahead and kill yourself?' "[1]

Most reactions to suicidal conduct are subtler but equally retaliatory.

The professional who permits the suicidal patient to carry out his intention of killing himself is committing a retaliatory act of counteraggression, a spiteful, vengeful response to the suicide's murderous hostility. The response is cloaked in the mantle of concern for the patient's civil rights, or disguised by other rationalizations, but this does not detract from its innate viciousness. Failure to act here, to prevent suicide, is a serious violation of professional ethics because it ignores clinical necessity—that is, the true needs of the patient.

Suicidal threats and suicidal acts are among the most anxiety-provoking, challenging, potentially devastating clinical problems with which we deal as health professionals. We cannot escape the responsibility of dealing with them actively and through primary prevention. In almost every case of suicide, there has been a cry for help—muted, distorted, hidden—at frequencies heard, often only by those trained to hear them and willing to hear them. We are obliged to heed this cry for help and to respond to it in ways that go beyond common sense. We must afford the patient another chance at life and at the psychological truth about himself that he has been blocked from understanding.

Note

1. Copyright 1981, Universal Press Syndicate. All rights reserved.

16 The Ethics of Suicide: Act and Intervention

Colleen D. Clements,
Roger C. Sider, and
Richard A. Perlmutter

A good deal of attention has been given to the possibility and ethical status of rational suicide in medical ethics literature.[1] This description of the problem of suicide derives from the concentration by traditional ethical theories on the cognitive aspects of human behavior—for example, on the characterization of man as primarily rational—and on the emphasis on free rational choice as the overarching value—autonomy. Construing man as the moral agent in these terms, however, is far removed from the complex world of human beings and human relationships because such a moral agent is asocial, ahistorical (both developmentally and culturally), nonphysical and nonpsychological. Within this framework, the relevant question concerning the ethics of suicide becomes the suicide act's rationality or irrationality. We propose a different approach: (1) that we can make no neutral judgment concerning the good of suicide and that the real ethical concern is rather the good of intervention and (2) that an analysis of the good of intervention requires a broader concept than that of the rational moral agent.

Our general ethical position is that values make sense only within an affirmative stance toward existence or experience and that they are not neutral, therefore, toward affirmation or negation.[2] Affirmation is a prerequisite for ethics, as it also is for the cognitive and aesthetic values. Suicide is a central issue in value theory, then, because as the acting out of negation, it forces us to look at the grounding for values and to see that grounding, not as a neutral Archimedean ethical point but as intimately involved in an affirming attitude. Any attempt to make an overall ethical evaluation of suicide (negation) is to beg the very question, since the evaluation is made from an affirmative perspective.

We need to be clear about our use of moral agency in discussing the affective character of choice (affirmative or negative attitude). We do not operate with the rational, asocial unit of moral choice in this description. A long evolutionary and historical process has brought us to a particular point

---

Colleen D. Clements was supported in part by a grant from the Ethics and Values in Science and Technology Program, National Science Foundation, Grant No. 055-8018092.

in valuation, and in ethics, as in other human endeavors, we stand with others on the shoulders of those who have preceded us. Those of us who affirm existence have a consensus and a tradition and do not choose alone. Our choices have the accumulated experience of human history that we may at times put aside or modify to fit the particular new situation but also that prevents an individual idiosyncrasy from becoming arbitrary. We function ethically within a matrix that is more descriptive of moral agency than the autonomous, rational unit agent of many philosophic positions.

It would be both radical and unjustified to dismiss the moral-agent concept, for the consequences would be profound for ethics, law, and society. We are not denying moral agency, however. To do so would be to make ethics impossible. Rather we are rejecting the philosophical definition of the moral agent as simply not representative of human reality. Clinical practice compels an alternative model because, in real-life situations, morally relevant decisions are deeply embedded in this matrix of conditioning influences. For suicidal persons this implies serious doubt about the frequency (or even possibility) of a totally free, autonomous suicide. (One of us, working in the clinical setting of an urban general hospital emergency department, has not seen a single such case in eight months of observation.)

The person who is intent on dying unambivalently (rare as he is), will find a way regardless of our actions. Long-term, reasoned, existential choice to suicide remains an individual's prerogative, amenable to only brief delays by rigorous intervention, restraints, and protection. Usually, however, the clinician sees suicidal persons who are driven by temporary impulses to kill themselves after a loss, who have become self-destructive secondary to an underlying state like psychosis or depression, or whose suicidal behavior occurs in the context of characterologic problems and/or family strife. In some cases suicide attempts almost become a way of life. Clinical practice, then, directs us toward a view of suicide that is more highly conditioned than free and of suicidal persons as more interdependent than autonomous.

Nonetheless, the question of therapeutic intervention to prevent suicide is one of the most enigmatic ethical issues in medicine. Health care professionals are trained to work on behalf of the patient's welfare in collaboration with the patient who, it is assumed, will be motivated toward the same end. It is only slowly and painfully that caregivers learn of the depth of the ambivalence with which some persons regard their health and even their life. Faced with such patients, clinicians experience not only a disquieting clinical dilemma but also a fundamental ethical conflict. And while many would be unable to articulate their problem in ethical terms, the conflict is clear. Suicidal patients constitute an ultimate threat to the doctor-patient relationship for their behavior denies that there can be any basis for collaboration or foundation for a therapeutic contract. Ethically, the clinician is caught between values that, although usually congruent, have become polarized. On the one hand is the clinician's obligation to secure the physical welfare of his patient. On the other is his respect for his patient's responsibility and liberty. In fact, the patient's consideration of suicide calls into question the working medical values of the profession specifically and calls into question the justification for usually unquestioned communal values in general. The values of well-functioning, of existing, of bodily repair, of psychological healing all, in spite of common human consensus and a professional tradition, now require some kind of justification, some grounding, before we can again be comfortable with them. The clinician, as representative of the values of all clinicians and the common human choice for life and function, must now consider those choices in the presence of someone who questions them. He must validate what the majority of humanity, what his profession, usually accepts without requiring validation. Suicide frames the ethical question quite clearly and fundamentally.

Moreover, efforts to provide for clinicians a calculus for action in such situations have been disappointing. As we have argued, philosophical contributions usually miss the point by considering the question from the standpoint of the person contemplating the act and seek to discover a method by which the act may or may not be justified. However, whether or not suicide is justifiable from the victim's perspective is of limited value to the clinician because the clinician is not primarily concerned with the ethical status of his patient's behavior. Indeed, a nonjudgmental stance is regarded as a clinical virtue. Rather, the clinician is concerned with his patient's behavior as it affects health achievement and health maintenance. (On this point, it has been argued that suicide can, in fact, constitute a healthful act, but this requires a sufficiently radical departure from customary values in medicine as to be intuitively unacceptable to most clinicians.)

What is required by the clinician, then, is not an ethics of suicide but an ethics of suicide intervention. Health care professionals need to know whether, when, how, and to what degree to intervene in cases of suicidal threat. Even in this more limited domain, conflicting advice is offered across the entire range of possible postures. On the one extreme of intervention is the orthodox psychoanalytic dogma that maintains that no suicide attempter really wishes to die—that all attempts are motivated unconsciously by aggressive impulses originally directed toward another but now retroflected against the self. Although this is no doubt a common dynamic, to universalize it is to dehumanize the patient. In denying any person such an option, and doing so on theoretical grounds, the therapist is depriving him of a distinguishing characteristic of humanity, the ability to reflect critically upon his own life. On the other extreme is the noninterventionist stance of Szasz—that there can be no justification for intervention for suicide prevention in any case.[3] This posture elevates the value of individual autonomy to preeminent status and simply denies the accumulated data—for example, the pervasive social influences upon cognition and behavior and the facticity of medical and psychiatric conditions that seriously circumscribe the domain of autonomous functioning.

The demands of an absolute calculus for action, which prescribes behavior for the clinician dealing with suicidal people, need to be replaced by a new way of thinking. We are suggesting a case-centered situational ethic that goes deeper than the question of whether or not to detain the suicidal person. A behavioral and dynamic understanding of each suicidal person we contact, and of our own motives and unconscious forces, must inform the ethical choice.

Conditioning Factors

We propose an ethical middle ground for intervention, but before doing so, it is necessary to describe the nature of conditioning factors that commonly impinge upon the lives of patients considering suicide. These factors can be broadly subsumed under the headings biological, psychological, and familial/social. It is, of course, obvious that such an artificial separation of categories for descriptive purposes does not convey the cohesive holism that comprises the human being.

Biological

The extent to which suicidal acts are biologically conditioned remains unclear. However, in two physical conditions, affective disorder and chronic physical illness, the risks of suicide are substantially higher than in the general population.[4] In the case of affective disorder, it is by now established that a genetic component is operative in some instances. This genetic component provides an antecedent conditioning effect, to be added later to whatever more proximate influences may be operative in the patient's life. Moreover, recent neurophysiological research is beginning to sketch mechanisms by which such genetic influences may be mediated. In the case of chronic, especially progressive, physical illness, the incidence of suicide is also elevated. Viewed from one perspective, this influence might be viewed as psychological since it appears to be related more to the patient's assignment of significance and meaning to the illness rather than to the illness itself. Yet we would argue that this is a biologically conditioned effect since it derives from somatic realities. To be alive is (at least in this life) to be embodied. Hence, our physical status may in itself alter the balance of life-affirming or life-denying tendencies.

Of course, the entire range of somatic influences upon mental states is enormously complex and, insofar as they induce a state of mind in which suicide appears attractive, they are relevant to our discussion. Clinical studies have demonstrated, for example, that cancer, brucellosis, some endocrinological disorders, and certain prescription drugs are correlated with an increased risk for depression. Clearly, then, the somatic status of a patient is always a potential source of conditioning influence regarding suicide and will, in some cases, be decisive.

Psychological

Dominant psychological theories regarding suicide are psychodynamic. Following Freud,[5] Menninger,[6] and later ego analysts,[7] psychodynamic theory posits various mechanisms in the genesis of self-destructive impulses and acts. Among them are retroflected aggression, passive death wishes, masochistic suffering, and the collapse of self-esteem secondary to deficient external supplies of psychological support and love. Moreover, at the level of character structure there is said to be a preponderance of oral, narcissistic, and dependent tendencies in patients who are chronically self-destructive. From a developmental perspective, there are many hypotheses but relatively little data regarding the impact of early-life experience, particularly deprivation, on later character structure and suicidal behavior. What is important here is the realization that suicidal behavior is the outcome of both conscious and unconscious processes in individual psychology, processes that are often irrational or nonrational and about which the suicidal person has limited understanding.

Familial/Social

Suicide in the context of family issues is the rule rather than the exception. Several authors,[8] in fact, see all suicide attempts as family related and state that treatment must address these issues. "The maintenance of the bond among members is therefore a life and death matter."[9] Observations and interviews with suicidal people and their families repeatedly show disturbed family relations, especially deficiencies in family members' capacity to empathize with the suicidal person. Especially prominent are severe conflicts over attempts of one member (usually the suicidal person) to develop an identity separate from the family's. The suicide attempt can be interpreted as a desperate effort to separate from, reunite with, or punish the clinging family member. The suicidal behavior is aimed at relieving painful conflict either by death or by evoking desired behavior from others. Other family determinants include a need to offer a victim as a way of distracting the family as a whole from other distressing problems. The suicide attempt or completion may occur in the setting of a family that subtly and unconsciously wishes the person to die. We often hear statements like "He can't even do that right"; "His life isn't worth living anyhow"; or "I guess we would be better off without him." To not intervene in a person's suicidal plans may be to sanction the family's hostile dynamics. It is an example in which not intervening couched in terms of siding with free will may actually be siding with death. In some cases, however, suicidal behavior occurs in the context of relatively benign family dynamics. In these cases the causal connection may be in the opposite direction—that is, the suicidal person's behavior can induce or exacerbate family disturbances.

At a broader social level is Durkheim's theory regarding the social origin of suicide—the anomic, altruistic, and egoistic types.[10] Here, suicide is seen as resulting from the nature and degree of the socializing process itself, which can not only reinforce a suicide choice but also can be the primary causal factor. While there are problems with the anomic concept, and while Durkheim really argues for an exclusive rather than inclusive social-level explanation of suicide, the social perspective is certainly one explanatory perspective. We would only like to caution that all perspectives have some claim to explanatory power.

Ethics of Intervention

Clearly, our emphasis upon the clinical complexity of conditioning factors in each individual case and our skepticism regarding philosophic abstractions leads us toward a case-centered ethic. We contend that this is not ethical relativism but is required for the clinician to work with integrity on behalf of his patients. As McConnell states, "It [case-centered ethics] is not only an alternative to the proscriptive, often prejudicial, and always indiscriminate applications of code ethics but an attempt to take seriously the empirical world and the data of personal relationships as crucial to moral decisions."[11] We will now describe our ethical approach in terms of whether to intervene and the nature of intervention.

Whether to Intervene

Whether to actively intervene in a case of threatened or attempted suicide is not, as we have stated, addressed by appeal to the concept of rational suicide. Rather, intervention can be justified only from within the perspective of the intervener who, it is assumed, has an affirmative stance toward life. Within that perspective we can assess the potential impact of the suicide upon ourselves and others in his social field. Frequently, such acts direct hostility or aggression toward others in the social network and may represent a serious loss. Our capacity to empathize and our feeling of compassion for those thus affected, as well as for thc suicide himself, can result in a great deal of pain to us if the suicide is completed. Moreover, suicide is profoundly disturbing and may represent a threat to our own affirmative attitude. The response to the Jonestown suicides indicates the severity of this threat on a large social scale. It must be admitted that this consequence is not a necessary one, as illustrated by the Stoics' use of the "open door" (possibility of exiting an unbearable reality), which served to maintain an affirmative feeling toward existence; but that was another time and another culture. Given these undesirable consequences to those making the decision to intervene or not, our general response will be based upon the balance of how we would feel about restricting someone's autonomy as against the assumption of psychological costs of his suicide to us. If, as we have argued, the ethical justification for intervention rests upon the sentiments of the intervener, we must address the issue of countertransference because it could be charged that such a foundation is nothing more than massive countertransference. If that were the case, the motivation for intervention would be neurotic rather than ethical. Although this is a serious question, it can be addressed along the following lines.

First, the term *countertransference* has come to have a whole range of meanings. Technically it denotes the analyst's unconscious emotional response to the patient's transference, or the analyst's emotional reactions to the patient, often only partly conscious, that are based upon earlier important relationships in the life of the analyst. In current usage, however, it denotes the entire range of emotions evoked by the patient in the therapist. In the latter case, since both reality-based as well as neurotic emotions are involved, the label countertransference is not itself prejudicial to our case. The feelings that justify suicide intervention are precisely *not* those that are technically countertransferential. Rather, they rest on those feelings common to all appropriately socialized human beings, of empathic caring for the other, that are the basis of all ethics.

Moreover, it has been established by now that the impassive objectivity of the analyst, the reflecting mirror allowing the patient to see himself, is not a comprehensive characterization of the helping relationship. Therapists, as all health practitioners, must bring medical/professional values into the clinical encounter, first to motivate their work with patients and, second, to give direction to their interventions.

Finally, the sorting out of ethically valid from invalid feelings toward the patient is a special case of the general problem at the heart of psychotherapeutic practice—namely, realizing when therapeutic impulses are motivated by legitimate relational feelings and when they are motivated by neurotic or inappropriate countertransference. This is a foundational aspect of professional training, and it is reasonable to expect that well-trained therapists can make this discrimination with some consistency.

We believe, then, that in grounding therapeutic intervention upon the therapist's own feelings, we need not build upon the shifting sands of neurotic countertransference. Rather we are describing an ethical perspective that is descriptive in fact of what clinicians do.

Second, within a broad biopsychosocial model we can assess the presence and magnitude of those conditioning factors that may render the suicide's self-destructive tendencies more compulsive than free, more conditioned than autonomous. Still, it can be argued that justification for intervention on these grounds requires an ethics of well-functioning that presupposes the affirmative attitude—thus, that it is again begging the question. It is true that yea- or nay-saying, while prerational, is not preexperiential and that the presence of sufficiently decisive conditioning influences may preempt the possibility of a choice at a mental level. Yet the value placed on choosing needs grounding; the answer to the question "Why value choosing?" depends on our finally valuing some minimum norm of human functioning, and well-functioning places us squarely within an affirmative stance. Why value well-functioning?—because it contributes to our happiness and survival. Why value pleasure and existence?—simply because we have an affirmative attitude toward the possibilities of human existence. This human consensus that we share with those in the affirmative stance toward life underlies our professional values but is precisely what is called into question by the suicide choice. The biopsychosocial model can give us justification for the timing, degree, and nature of intervention (a means of reality testing) but not a justification for whether to intervene or not.

In our scheme, then, there is no independent arbiter who rules on the ethical status of both the suicidal person's and the care giver's behavior. And the care giver is left to justify his choice of intervention from within his own commonly shared affirmative stance (and its long, accumulated tradition) rather than from within the point of view of the suicidal person. There is assistance with this choice. It is not the isolated therapist as lone moral agent; rather, it is the professionally integrated therapist (with the long tradition of medical values and accumulated categories for explaining experience backed by pragmatic experience with applying those values and categories) who makes a choice that is in many senses collective and therefore not singularly arbitrary.

Other traditions can be of assistance, one of which would provide an overarching perspective from which the acts of both parties could be evaluated—that is, the consideration of the question from within a theological framework (which, of course, would also need some sort of justification). Here, morality finds its basis in a transcendent deity, God, and is construed in terms of the sacredness of man in the natural order and the covenantal character of man's obligations to his creator and to others. For the special case of those persons affirming the tradition and the content of the Judaeo-Christian world view, both suicidal patient and therapist are members of a common community of faith. The dialogue concerning the ethics of suicide can then take on a considerably altered character. Specifically, it may provide a rationale and, in some cases, content and method for intervention. Although this grounding of ethical behavior is unavailable to many persons at the present time, its adequacy for believers is easily demonstrable.

The Nature of Intervention

The ethical approach of the clinician to the suicidal person must begin with empathy for it is the basis for hope and the sine qua non for problem solving. The therapist who can reflect just how hopeless the person feels ironically makes an intervention since the person feels that he is being understood rather than coerced. The empathic therapist avoids the trap of viewing intervention as another conditioning force, pressuring someone to live as he is or arguing that his life is worthwhile. Such an approach is ineffective and often counterproductive. Rather, the human encounter that allows therapist and patient to study the dilemma and issues involved in negation can by itself be helpful and decrease the tendency toward compulsive expression of interpersonal anger. Thus, empathic understanding allows the beginning of a sorting-out process to determine whether the suicidal stance truly reflects a choice to die or, as is much more common, a step of desperation in the face of overwhelming psychic pain, insoluble circumstances, and chronically unmet personal needs. Another coordinate of clinical intervention concerns the qualitative nature and extent of protective measures. Discussions around intervention with a suicidal person often revolve around the ethics of incarcerating persons against their will versus allowing them to exercise their right to die by their own hand. It is important to note the oversimplified nature of this dichotomization. Suicide-prevention measures are not ethically homogeneous but rather derive their character from the nature of the intervention. For many acutely suicidal people, hospitalization can be helpful, often lifesaving, and yet, in clinical emergency psychiatric practice, a group of patients are identified who are habitually self-destructive and even suicidal. They are known by many names including borderline characters, suicidal characters, repeaters, self-mutilators, and more. Although the intent behind the behavior is more for human contact than for death, the life-style becomes high risk. This poses a common and difficult ethical dilemma: Can you allow someone who is self-destructive to return home?

Admission to a hospital for those patients is fraught with dangers of regression, anger, and even more serious self-destructive behavior. Yet the presentation is worded something like, "I took the pills (or knife) because I wanted my lover back, and now if you don't admit me I will kill myself." If hospitalization is agreed to, the risk is that the person will be reinforced in the knowledge that emotional pain and life problems can be handled (or avoided) by attempts to kill or mutilate himself. In the future, the person more easily may choose life-threatening behavior as a desperate measure to get hospitalized. Reflexively, hospitalizing such suicidal people may inadvertently set up a deadly pattern. For example, the patient who persistently argues for death (nay-saying) then becomes dependent on the therapist who argues for life (yea-saying). One day the therapist may be unable to engage in this struggle to preserve life, and the patient who has not taken the struggle into himself becomes seriously suicidal.

From this standpoint it can at times be ethical not to hospitalize a patient who says he will kill himself if you send him out. A more reasoned ethical intervention might involve keeping responsibility for the life-or-death decision on the patient, working to restore an active outpatient regime to help the person deal with intolerable affect and impulses, and offering consistent empathic contacts while the person is in pain and crisis. This illustrates how not hospitalizing a potentially self-destructive person may, in fact, represent the yea-saying stance.

Ultimately, clinicians act in ways they believe will be relatively effective at a cost that they are prepared to bear. Moreover, they bear in mind that suicide precautions are never completely effective and that they do not make successful suicide an absolute impossibility. Rather, therapeutic interventions are calibrated to substantially reduce the probability of a fatal outcome.

This double character of clinical interventions—that they simply alter the odds and that they carry costs—is ethically relevant because it illustrates the fluid and ambiguous character of intervention. When have the risks been reduced enough, and what cost is acceptable to the patient, the clinician, the family, and society? These are questions that do not yield to codes or abstract analysis. Yet the clinician computes these variables daily. In general, he attempts to reduce risk as much as the situation requires and at a cost that is deemed acceptable.

Intervention in the form of forced commitment is sometimes described as a deprivation of freedom, choice, or autonomy. This is not an adequate account. Rather, attempts to prevent or delay suicidal behavior are attempts to help regain the freedom that the person has lost by virtue of treatable psychiatric disorders, temporary overwhelming loss or stress, or relations with others. Such people have lost the judgment to make an irreversible decision. If we assume, as Eichelman states, that people are not expendable, then physical protection, even if involuntary, can be an act of the highest respect.[12]

The clinical matrix remains affirmation. The intervention will often involve physical protection from self-destructive impulses, but as we have discussed, protection is not the sole ethical consideration. The ethical question shifts from whether to intervene to how best to design the intervention. It may involve helping the person become more aware of alternatives, of choice, of his power. It may be designed to improve the quality of life or treat an underlying disorder. The intervention may aim at changing a cognition around a negative self-image or restoring a human relationship.

A third coordinate is one of time. Suicidal risk is the sum of life-affirming and life-denying processes that are dynamic and in constant flux. An intervention is not an end in itself but rather looks forward to the time when risk will be moderated so that intervention is no longer required. Thus, intervention is intrinsically a temporizing process. To intervene on behalf of a suicidal patient is to bide time. It is hoped, of course, that doing so will allow the changes to occur that will diminish risk—the amelioration of a depression, an improvement of physical circumstances, the establishment of meaningful relationships, or perhaps an altered view of life and its meaning. The clinician is attempting to protect the patient over an acutely self-destructive period until he is once again able to affirm life.

In ethical terms, this allows the physician to pursue the patient's welfare in violation of the patient's autonomy only for a time-limited period. If the clinician cannot realistically anticipate sufficient grounds for the patient to reverse himself within a reasonable period of time, or if the patient in fact does not do so after a suitable period of treatment even though he might have been expected to, the clinician's ethical obligation becomes progressively more tenuous. Thus, a therapeutic intervention is readily justified in situations of acute suicidal risk, but when that risk becomes progressively extended across time (for example, some borderline patients), intervention becomes more problematic and cannot be viewed simply as an obligation to protect. Although there is no formula for this attenuation of responsibility, its temporizing nature is clear. To assume perpetual responsibility for physical protection of a suicidal patient is not to treat him but rather to infantilize and dehumanize him.

Conclusion

What we have been developing, then, is a very different perspective on the ethics of suicide based on what we see as most adequate theoretical and clinical grounds. Theoretically, in terms of a general axiological (value) theory, to apply the concepts good or bad to the suicide act or to the person considering the choice makes no sense. Such concepts depend on an affirmation of existence for the meaning they have so that suicide *is* truly the only foundational ethical question. It forces us to ask on what our values are based, and should help us understand that they are based on the values of existing. To assume we can evaluate suicide *for* the person considering it, with the tools that assume an antisuicide attitude on our part, is to beg the question on a grand scale. Clinically, we usually shy away from such judgments concerning our patients' values or choices, but suicide choices are more problematic because, again, they are a foundational problem for the value of medicine, which assumes primary goods of well-functioning or existing, the grounds for which are called into question by suicidal ideation. The same question-begging problem arises.

We need to shift our focus, then, from the meaningless search for an ethical statement about suicide (unless from a theological perspective) to the ethical justification for intervention based, not on the intrinsic nature of the suicide act or choice but on the needs and interests of an affirming human community. This seems at first a radical move, but we would argue that it most accurately describes the situation. We choose to intervene because of values *we* hold about well-functioning, existence, and the potentials for human life and because, as empathic, social beings, we feel for others and are motivated by that feeling. We can anticipate the collective pain to others that suicide can produce and can choose to accept that pain generally under unusual circumstances only (for example, some cases of terminal illness or repeated therapeutic failures). Thus, the ethics of suicide needs to be centered in the values of those in an affirmative stance, and the justification for intervention is that the interveners have their own good reasons.

Notes

1. M. Battin and D. Mayo, eds., *Suicide: The Philosophical Issues* (New York: St. Martin's Press, 1980); and T. Beauchamp and S. Perlin, eds., *Ethical Issues in Death and Dying* (Englewood Cliffs, N.J.: Prentice-Hall, 1978), especially chapter 2.

2. C. Clements, "The Ethics of Not-Being: Individual Options for Suicide," in *Suicide: The Philosophical Issues,* edited by M. Battin and D. Mayo, pp. 104–114 (New York: St. Martin's Press, 1980).

3. T. Szasz, "The Ethics of Suicide," in *Suicide: The Philosophical Issues,* edited by M. Battin and D. Mayo, pp. 185–198 (New York: St. Martin's Press, 1980).

4. H.L.P. Resnick, "Psychiatric Emergencies" in *Modern Synopsis of Comprehensive Textbook of Psychiatry,* vol. 2, edited by A.M. Freedman, H.I. Kaplan and B.J. Sadock, pp. 2085–2098 (Baltimore: Williams and Wilkins, 1980).

5. Sigmund Freud, "Mourning and Melancholia," in *Collected Papers,* vol. 4 (London: Institute of Psychoanalysis and Hogarth Press, 1924).

6. K. Menninger, *Man against Himself* (New York: Harcourt, Brace & World, 1938).

7. E. Bibring, "The Mechanism of Depression," in *Affective Disorders: Psychoanalytic Contribution to Their Study,* edited by P. Greenacre, pp. 13–48. (New York: International Universities Press, 1968).

8. K.S. Adam and G. Adam, "Attachment Theory and Attempted Suicide," (Presentation at Royal Australian and New Zealand College of Psychiatrists, Singapore, 1978); and J. Richman and M. Rosenbaum, "A Clinical Study of the Role of Hostility and Death Wishes by the Family and Society in Suicidal Attempts," *Israel Annals of Psychiatry and Related Disciplines* 8 (1970):213–231.

9. J. Richman, "The Family Therapy of Attempted Suicide," *Family Process* 18 (1979):131–142.

10. E. Durkheim, *Suicide* (New York: Free Press, 1951).

11. T.A. McConnell, "Suicide Ethics in Cross-Disciplinary Perspective," *Journal of Religion and Health* 7 (1968):7.

12. B. Eichelman, "Ethical Aspects of Psychiatry," in *Psychiatry and Behavioral Sciences,* edited by D. Hamburg, J.D. Barchas, P.A. Berger, and G.R. Elliott (New York: Oxford University Press, in press).

17 The New Look of Death

Daniel C. Maguire

The age in which we live has been dubbed the age of death. Social psychologist Edwin Schneidman writes: "In the Western world we are probably more death-oriented today than we have been since the days of the Black Plague in the 14th Century."[1] In the United States, the obsession with death is in full bloom. This is to be welcomed. It is not a morbidity boom but a coming of age, a rite of passage. U.S. culture is adolescent and spoiled. We have not grown up with limit. Death, Arnold Toynbee chided us years ago, is viewed in this country as an un-American activity. Yet two million Americans a year indulge in it anyhow, so it is a good thing that we have begun to take notice of our mortality. As far as we can tell, we are the only animal graced with a consciousness of our mortality. As soon as we come to know with any maturity, we come to know that death is our portion. A culture such as ours—that has hidden from death, that encourages people at wakes to comment on how good the corpse looks, that chemically fortifies the bereaved so that they will not disturb our funeral-denial process with displays of grief—has problems that go beyond death to life.

Thus, it is good for us that death in our day is having its belated due. A revolution of consciousness in this area will have consequences that go well beyond questions of mortality. In our lives there are certain ultimates—for example, death, sex, religion, good, and evil. It is also clear that when a shift in consciousness occurs about any ultimate reality in the human ambience, the effects are seismic and widespread. Critical reflection is important at a time like this, when some of the foundations of consciousness are shaken. It is to be hoped, since we all must do it, that in this time we will all learn how to die. Maybe soon we will be able to plumb the meaning of some of the ancient sayings that have long been most foreign to the way of death in the United States. *Bene mori,* "to die well," was an ancient ideal. Similarly the ancients could speak of the *ars moriendi,* the art of dying. These expressions jar a death-denying culture. We are offended by the wisdom they suggest.

In this chapter, therefore, I will explore the manifestations and effects of the changes in our death consciousness. First, I will note the changes in our relationship to death itself and to our moral dominion over the dying process. Then I will look at the fallout in other related areas affected by the profound shifts in our mortality awareness.

There is an old Irish saying, relayed to me many times by my mother, that "there are worse things than dying." There is simple truth in that, and yet it seems that the simplest of truths are the most difficult to penetrate. We have seemed to assume that life in any condition is preferable to death, and that is simply not self-evident. Hilaire Belloc chided this attitude years ago with this ironic verse:

> Of old when men lay sick and sorely tried
> The doctor gave them physic and they died.
> But here's a happier age for now we know
> Both how to make men sick and keep them so!

We need mention only the name Karen Ann Quinlan to know that this is not a happier age in this regard. The life signs of Karen Ann have been preserved for years now at a cost of millions of dollars in legal and medical expenses and at the cost of infinite misery for her loved ones. Yet her brain is so massively damaged that she will never return to conscious life. There are worse things than dying, and Karen Ann's treatment shows that to be so; she has been denied her right to a good death.

Years ago, of course, there could be no Karen Ann Quinlan case. We could not have provided the technology needed to keep her life signs going. What we have done is to acquire the ability to produce what one physician called the half-cure. We can keep a person propped up half-dead and half-alive, when years ago they would have died and people would have called the death a blessing. Pneumonia used to be referred to as an old person's friend. The idea behind this statement was that when a person was dying in great misery and pain, pneumonia often cut it short. A strong heart might have kept them going longer if pneumonia had not entered with its friendly interference. Now we can enter in and, with antibiotics, prevent the pneumonia from rendering its friendly service. Is this progress? It is a blessing that we have such control over pneumonia, but is the use of this power always a blessing?

If we look carefully at the saying that pneumonia is an old person's friend, were we really saying that pneumonia was the friend or were we saying that the pneumonia gave access to the last remaining friend, which was death? Death must always be considered an enemy until it shows itself to be a potential friend. Life is good and is, indeed, the precondition of all the good we experience. Yet the life that is good also bears the mark of the tragic. There are times when the end of life is the best that life offers. To know that is to know with the Irish that there are worse things than dying; to know that is to be wiser.

Doctors, of course, are not always to be blamed for the unreasonable prolongation of life—which amounts to a prolongation of dying. Doctors are under legal, personal, and professional pressures. Most states do not protect doctors from the dilemma they face when they are sure the respirator will not restore the person to conscious health but are afraid to turn it off and let the patient die. For one thing, the doctor could be criminally liable if a district attorney decided the doctor took a positive step (turning off a machine) and death resulted. He could also be judged as having violated his healing contract. Even if the district attorney avoided such witch hunting, a distressed relative could sue and tie up the doctor's time and money for some time.

No, the culprit cause is not isolated in any one profession. It lies in our deeper cultural uneasiness with death. Needed in every state is what could best be called an appropriate-care act. (The Natural Death Act, following the California example, has been used by the states that have such legislation, but this term is less descriptive of legislative intent.) An appropriate-care act would stipulate that we have an unqualified obligation to give appropriate care to the sick and the dying. (Indeed, we humans believe usually that appropriate care should be taken even of the remains of the dead.) Clearly, while there is life there should be appropriate care. Our duty is to company with the dying as they meet that special challenge that awaits us all. At the beginning, it might be appropriate care to do very dramatic things for the patient, including heart by-pass surgery, the use of respirators, and so on. However, if our best efforts there do not return the patient to some modicum of conscious health, it may be that the nature of what care is appropriate will change. Later in the illness, when it is clear that dramatic interventions of medicine will not yield the desired result, it may be appropriate only to hold a hand, wash a brow, adjust a pillow, whisper prayers in an ear, and let the dying die. Precautions that premature dismissal of patients would not take place could be built into such legislation, as the existing legislation shows. With such laws, the ideal of the good death could be realized.

Other, more serious questions have arisen in this time regarding our moral dominion over the dying process. Stuart Alsop, while dying, wrote about a hospital roommate of his. He described how the man, dying in terrible pain, would receive his dosage of painkiller. He would then sleep quietly for an hour. During the second hour, the pain would begin to return, and the man would begin to whimper quietly. During the third hour, the pain would become more pronounced and the man would moan and groan in his bed. During the fourth hour, before his next dose of relief, the man, in Alsop's phrase, "howled like a dog." Many reactions are possible to this. Some physicians will say that with good pharmaceutical care such pain is unnecessary, but few physicians will say we have conquered all pain. Others will say that the man, if he had requested it, should have been given whatever was necessary to stop this agony, even if some dose might cause his death. This would be a conscious, direct act to relieve pain even if it causes death. The death that could occur would be partially willed because of the decision to give pain relief the priority over life prolongation. If you decide to relieve the pain even if it causes death, you have willed death into your painkilling project. Simply stated, that is mercy death, or what has been confusingly called euthanasia—a term that has too many usages to be useful.[2]

Could such an action ever be countenanced? Would this be playing God or simply being responsible as a cocreator and coprovider with God, in the terms of Christian theology? Would this open up a floodgate of mercy deaths, or is it a rare situation that could be confined within moral limits? Serious theological questions arise here also. Ethics is serious business because ethical errors may not be just unfortunate but can even be lethal. Beyond that seriousness is the fact that, when we make an ethical judgment, we paint a picture of our God. When believers say that X or Y is moral or immoral, they have also said that this ultimately is the way they think God would see it. Thus, if you say that it could not be moral ever to hasten death, even in the tragic case under discussion, you have said that such is the mind of God. If, as many would say—and I include myself in that number—that such extraordinary interventions may in tragic circumstances be justified, that gives another image of what God would seem to be. God is remaining silent on the debate, and in that awesome silence we must make our moral decisions, firm in our belief that the mark of His/Her wisdom is on our minds since we were fashioned in the image of the divine.

Another important result of the new interest and openness to death is the hospice movement. The hospice movement is the focal point of the new attention on where one should best die. More and more in recent years, persons have been dying in hospitals and other institutions, which is anything but ideal. The atmosphere may be sterile and cold. Also, at a certain point, the dying person may actually be out of place in a hospital. A hospital is geared to healing, to graduating persons back into healthy life. When such is clearly not one's lot, and when the hospital has exhausted its healing efforts to no avail, that may no longer be the place to stay and die. Home may be better. To be with one's family in one's familiar setting at such a serious time is clearly ideal. Of course, the need for pain control and other scientific interventions might make the return home unfeasible; and that is when the hospice enters. Some hospice centers provide home care so that the patient can stay there as long as possible or right up to the moment of death. Others offer a setting of comfort and dignity, with opportunities for unlimited family attendance and first-rate pain management and counseling as needed. It is difficult to believe that any community would be lacking in such a humanizing care facility.

Another result of our new death awareness is a greater focus on the aging process. Gerontology, of course, is a new phenomenon. It is estimated that 25 percent of all the persons who ever reached 65 years of age are alive today. Since not as many persons lived into old age, less attention was paid in the past to the special problems of the aged. That is changing rapidly, however. Medicine is becoming more sensitive to the needs of the aged. Some nauseating discomfort that to a younger person might indicate either overeating or temporary nervous upset could, for the older person, signal the onset of death that, in advanced age, could come from so many causes. There may be a special panic to the sickness of the aged that was too little sensed in the past. Also, the tendency for a youth-worshipping adolescent culture such as ours has been is to undervalue the aged. In many cultures, the older a person was the more esteemed he or she was. The aged had tasted more of life and its surprises and were deemed wiser for that. This has not been the recent experience in the United States. To carve out this nation from the wilderness, youth and physical strength were prime credentials. In such an atmosphere, the prerogatives of age fell into understandable, but not forgivable, neglect. This long-tenured bias is now coming under scrutiny, and it augurs well for all of us who are embarked upon that process that in early life is called growing but that later is called aging.

Another important effect of the new stress on death and dying is a new appreciation of the phenomenon of grief and bereavement. Studies are now showing that bereavement can be a fatal disease. Drs. Dewi Rees and Sylvia Lutkins did a study of bereaved persons in Llandidloes, Wales, over a six-year period. The death rate among widowers and widows during the first year after bereavement was 12 percent as compared to 1.2 percent of their nonbereaved counterparts.[3] This has come to the attention of pastoral counselors and ministers. Bereavement calls for community because it cannot be borne well alone.

The impact of bereavement has led to other considerations of the impact of psychological trauma on physical health. Could serious personal shocks and sufferings cause or contribute to the development of physically pathological conditions? This question is no longer offensive to medical orthodoxy. Prior to this time, medicine seemed to work on the assumption that stress and mental suffering could have a physical impact in only two ways: headache and ulcer. Beyond that the patient was seen, or so it would seem, as a biochemical preparation beyond the reach of the psyche. A number of studies are now questioning that. One study, for example, considered a number of women who were suspected of cancer of the cervix. While these women were in the hospital for a biopsy, the investigating physicians interviewed them in advance seeking to see whether they had suffered a serious break in significant social relationships and whether they had reacted to the biopsy with feelings of hopelessness and depression. If they had, the investigators predicted cancer and were correct in 75 percent of their predictions, both positive and negative.[4]

Obviously this research is dangerous in the sense that the possibility of hidden variables looms large. Nevertheless, there does seem to be here a raising of consciousness in an area of supreme importance, an area that was heretofore blocked by untested orthodoxies.

Another development related to the new interest in death is in the health care field. Having looked at the dying, we look also to the ill and to those who attend them. The role of the physician is being demythologized. A recent change in the name of a hospital in Milwaukee may serve as a symbol of what is going on; the hospital that used to be called Doctors' Hospital changed a few years ago to Family Hospital. The doctor, in many ways, has been a high priest in our society, which is not surprising. The aura of the sacred attaches to experiences of the ultimate, and the doctor presides over ultimate experiences—birth, death, and healing. Thus, the doctor's role takes on sacral hues, and almost mythical powers are attributed to him or her. This was not a healthy situation. The sacral myth has been a distraction from the reality of the doctor as an applied scientist working with limited skills in the imprecise science of medicine. Under cover of the myth, incompetence and malpractice were protected. Under scrutiny, however, the myth is dissolving. Doctors in all their fallibility and humanity are appearing as they are. One unhappy effect of this is the epidemic of malpractice suits. The happy effects are that now we might enter as patients into a partnership with our physicians, not as passive objects but as codoers of health care. In the new demythologizing health care arena, we can put aside the infant-to-parent relationship with the physician and adopt an amateur-expert to trained-expert relationship. In the complexities of modern life, we must become amateur experts in many areas where ignorance would be too expensive. Certainly health care should be an area where amateur expertise would be a natural achievement of everyone. We could hope that even the nomenclature of the diseases and their cures will become less hostile and obscurantist. *Hepatosplenomegaly* is a good term for an enlarged liver and spleen if you are Greek. In English, the term *enlarged liver and spleen* would do nicely. George Bernard Shaw once said that every profession is a conspiracy against the laity, and the conspiracy against the medical layfolk has included a generous amount of obfuscation in terminology. Perhaps as doctors become incarnate with us as our brothers and sisters in health care, they will learn to speak a language that intelligent people can also understand.

The role of the nurse is also being reassessed. Nurses, traditionally being women, suffered from the antiwoman myth that is the deepest prejudice in the human species. Aside from that, the elevation of the doctor put down the nurse. The nurse was not able to grow as an adjunct physician with an augmenting and ever more creative role to play in health care. Changes here have begun, however.

A number of other issues, theological in nature, are coming out of the revolution in death consciousness. The meaning of suffering has come under reexamination. A sado-masochistic myth has long beclouded Christian understanding of suffering. Suffering was thought to have an intrinsic value. One had to offer it up, and it was transformed like a signed check into spiritual currency. Somehow it made reparation to God and squared debts between creature and creator. Now there is the chance to see that a genuine Christian view of suffering would recognize that, of itself, suffering is a physical evil, a manifestation of the incompleteness of creation. Those scientists who work to control it are doing God's work. Fabry's disease, which afflicts children, involves an enzyme deficiency. Recently, a cure has been developed to treat that disease by introducing the missing enzyme into the child's system. Screams turn to smiles and peace. The scientists who pioneered in this effort were, in religious terms, cocreators with God as well as coredeemers. Healthy Christianity can find no value in suffering in itself. The sufferer can bring value out of suffering. The value of the crucifixion was not in the cross or the crucifying but in the Crucified who brought meaning to it. The suffering of Helen Keller was transformed as she broke out of that terrible darkness of no sight and no sound and communicated with us and gave us all hope. The value was in the sufferer's ability to transform the suffering, not in the suffering. This is Christian and healthy and, in the current atmosphere of reexamination, more easily appreciated.

There is a new swell of interest also in the subject of the afterlife. Having become interested in death, it was but one more step to ask: Then what? An enriching literature on this long-neglected subject in theology is now emerging. This too we owe to the new look at death on which our culture has embarked.

Any shift in consciousness is frightening. Old patterns of thought are threatened and shaken. New ideas seem to profane old and cherished beliefs; but such times, too, are heavy with grace. People who walked in darkness begin to see great glimmerings of light. There are two kinds of people: those who fear the future and those who welcome it. Our future clearly involves a new penetration of the meaning of death. We should welcome it; the yield will also be a new understanding of life.

Notes

1. Edwin Schneidman, "The Enemy," *Psychology Today,* August 1970, p. 37.

2. In what was called the principle of double effect, earlier Catholic moralists urged that you could will the relief of pain and merely permit for proportionate reasons the unwilled effect of death that might result from the painkiller. This, in real life, is often the route to fruitless hairsplitting. In the case here, I am suggesting that the party would be accepting death as a directly willed risk. It would be artificial to say that death would merely be permitted but in no sense willed when you depart from the confines of supposedly safe dosages.

3. See W. Dewi Rees and Sylvia G. Lutkins, ''Mortality of Bereavement,'' *British Medical Journal,* October 1967, pp. 13–16.

4. Reported in Betty R. Green and Donald P. Irish, eds., *Death Education: Preparation for Living* (Cambridge, Mass.: Schenkman Publishing Co., 1971), p. 20.

18 The Hospice Movement: Science and the Terminally Ill

Dorothy H. Summers

At first glance, this chapter's title seemed to me something of a contradiction in terms. Hospices are specifically concerned with care at a time in life when aggressive curative measures using mutilating surgery, radiotherapy and cytotoxic therapy are no longer appropriate. However, when I looked up the meaning of the word *science* in the *Oxford English Dictionary,* it was defined as "an organised body of knowledge that has been accumulated on a subject." This definition really excited me, and my mind moved from the test-tube world of the science laboratory at school, past the machinery of modern technology, to the patient and his family, as this is the unit of care to which hospices must direct their expertise.

When St. Christopher's Hospice opened almost fourteen years ago, care of the dying had already progressed from the tender loving care of the medieval hospices where travelers rested on their journeys to the Holy Land. Doctors and nurses had already recognized that people who are dying are not just the failures of the curative system and that caring for them was and is a respectable part of medicine; difficult decisions have to be made, and considerable expertise and knowledge is necessary if symptoms are to be adequately controlled and a good quality of life achieved for both the person who is ill and for his family. These decisions should involve all members of the caring team—doctors, nurses, social workers, clergy, and ancillary staff—and our teachers are the patients, if only we will listen to them. Whose life is it anyway? is a question we do well to ask ourselves.

We at St. Christopher's are in many ways fortunate that, when patients are referred to us, the difficult decision has already been made that control of the cancer is no longer possible. This does not, however, exonerate us from the responsibility to continually review each individual situation when deciding what to do next. Our science is directed toward symptom control, and when this is achieved, the patient and his family will enjoy a quality of life that is not only worth having but also will leave families in bereavement with positive memories of the last weeks together.

Let us now focus on the principles of symptom relief and explore some of the factors contributing to the severity of symptoms. The symptoms of terminal cancer are chronic; this means that they give us no information we do not already know, they are always present as their cause is still there,

Figure 18–1. St. Christopher's Hospice in London

and they will tend to increase in severity as the disease progresses. There are also mental, social, and sometimes spiritual facets of the physical problem, and if total relief is to be achieved, these must also be considered. Intercurrent conditions must also be recognized and treated appropriately, and there is much detective work to be done—Sherlock Holmes, the famous English detective, could not use his famous comment, "Elementary, my dear Watson," to his friend Dr. Watson with any truth in these circumstances.

Let us use as our example the symptom of pain, since this is one most feared by patients, for many of whom cancer and pain are synonymous, though in fact only 50 percent of cancers are painful. This will enable us to explore the use of analgesics, and especially the use of narcotic analgesics, that are the subject of extensive research and yet which many doctors and nurses still fear to use and fail to understand. Chronic pain, already defined as useless, constant, and progressive, needs to be continuously controlled. A patient described this admirably: "Here you do not chase away the pain, you anticipate it," he commented as he was given his regular four-hourly medication.

When a patient presents us with a pain problem, the doctor must first listen to the patient's description, remembering that pain is what the patient says it is and not what we think it ought to be. Having done this, the doctor will prescribe an analgesic that he assesses will be adequate in potency and dosage to control this person's pain when given every four hours "by the clock." The nurse will be the person to observe whether or not this is effective, and if in twenty-four hours relief has not been achieved, the situation will again be reviewed. Many patients will eventually need a narcotic analgesic, but as long as simpler preparations are as effective, they will always be used. As a result of five years of controlled clinical trials at St. Christopher's by Dr. Robert Twycross, examining the comparative value of morphine and diamorphine and of the other components of the Brompton mixture, several interesting facts emerged:

> Given in individually optimized doses with a phenothiazine drug and by mouth, there is no clinically observable difference in the effects of morphine and diamorphine.
>
> Cocaine, which was originally in the mixture when it was used for the very sore throat of patients dying of tuberculosis and which was retained for its mood-stimulating effect, has little effect after five to seven days.
>
> The alcohol component—gin, whiskey, or brandy—does not affect the relief of pain.

The regime now in use when severe pain needs narcotics for its control is to give oral morphine in chloroform water every four hours, in a dosage from 2.5 mgms to 60 mgms, together with a phenothiazine drug—either prochlorperazine 5 mgms as an antiemetic or chlorpromazine 12.5 to 25 mgms if more sedation is needed for an anxious patient. The phenothiazine drugs also potentiate the action of morphine. When oral medication is no longer possible, we give by injection half the dose of oral morphine with a phenothiazine drug, and this is continued four-hourly. Other adjuvant therapy may be appropriate, and the use of steroids for inflammatory pain and antiarthritic drugs for bone pain will be most effective.

Having coped with physical pain, then, we need to look at the other contributing components that influence the patient's experience of pain. A fearful, anxious patient will often complain of excessive physical pain but does not find it easy to express his anxieties. We all have to learn to be good listeners; sitting close to patients and in silence that only the patient should be allowed to break can be a threatening experience for us, but it is invaluable to patients who are trying to find the courage to express their fears. Accepting death as being part of life, allowing dying patients to remain in the ward and not pushing the bed into a side room, and letting others see death as peaceful and that we shall be with them can be very helpful.

However, we must never force information, but rather allow patients to ask their own questions and choose their own confidante, knowing that they will always be given a truthful answer. After sitting with a patient, it is always important to look back as we leave the bedside—a changed expression can reveal a lot of unfinished business we have yet to tackle.

Social pain is perhaps the most difficult problem, for whatever we do we cannot alter the fact that the dying person has to part from those he loves; the patient thus experiences bereavement as well as his family. Perhaps the greatest stress is the conspiracy of silence that so often exists—patient's and family's finding it impossible to face the reality of the future and to share it together. Sometimes, one of us can help them to come close to each other, and if the dying person can join in decision making for the future of the family, this can result in great tranquility and a feeling of security about the future.

Spiritual pain is often shown by the patient's question, "Why should this happen to me?"; glib answers given with too rapid a response are quite unhelpful. To share the patient's feelings and show understanding of his questioning, but with the promise of companionship and prayerful support to the end, will often help patients through to acceptance.

One of the greatest securities a hospice can offer is that the patient and family will never be turned away and that the staff will journey with them to the end of the patient's life and on into bereavement for the family. Those in need of support can be identified, visited at home, and invited back to the Pilgrim Club—a social club for bereaved people. Research by Dr. Colin Murray Parkes has shown that this pattern of family care minimizes suicide risk, illness, and death during the first year of bereavement and brings families through the long valley to finding themselves again.

Science has brought us this far in caring for dying patients and their families, but the story continues and is far from complete. Those of us dedicated to this work continually have to listen to those who come to us to find many different ways of helping them, appropriate for each individual patient and family, and to enable each one to build a bridge of faith across the chasm of darkness to the light of a better tomorrow.

19 The Issue of Competency in Life-Prolonging Treatments

Paul Millard Hardy

The history of Western medicine over the past 300 years has been that of bringing the power of the scientific method to bear upon human disease and suffering. As a result, powerful new technologies have emerged in our lifetime that have become double-edged swords, having both the power to heal and the power to harm. These technologies have begun to raise many unique and profound problems for medicine since they represent scientific breakthroughs that move beyond the social, psychological, legal, and ethical capacities and constraints of our modern society. They have called into question many of our theological and ethical presuppositions, forcing us to reexamine what have been held to be most important. Our heightened control of nature requires us to reexamine the ethical and legal principles that we use to guide and motivate our actions in order to fulfill our moral and legal responsibilities.

Ethical reflection by physicians on the suffering of the incurable and care for the dying has been appallingly meager in U.S. and British medical journals from the nineteenth to mid-twentieth century. John Ferriar wrote in 1816 that no subject was less studied than the care of the terminally ill patient: "The Wise Look Beyond It, and the Inconsiderate Escape from [It]."[1] This neglect and associated suffering is perhaps best documented in Hector Berlioz's memoirs when he writes of the despair he felt when physicians refused to end the life of his dying sister:

> I have lost my eldest sister: She died of a cancer of the breast after six months of horrible suffering which drew heart-rending screams from her day and night. My other sister, who went to Grenoble to nurse her, and who did not leave her till the end, all but died from the fatigue and the painful impressions caused by this slow agony. And not a doctor dared to have the humanity to put an end to this martyrdom by making my sister inhale a bottle of chloroform. . . . The most horrible thing in the world for us, living and sentient beings, is inexorable suffering, pain without any possible compensation when it has reached this degree of intensity; and one must be barbarous, or stupid, or both at once, not to use the sure and easy means now at our disposal to bring it to an end. Savages are more intelligent and more humane.[2]

The despair and bitterness on the part of family members persist to this day as physicians, the courts, and moralists have failed to reach a consensus on how to care for the terminally ill in light of our advancing technology. One only has to look back to recent events to find just as poignant an example of bitterness and despair. Robert Spring, the son of the 78-year-old demented man who was maintained on dialysis for approximately fourteen months after the family and his physician felt it appropriate to withdraw this life-support system, spoke of his father's case two days after he had died a so-called natural death as follows:

> I remember when I was a small boy. We had a chicken farm. We used to throw the dead chickens into the field and plow them under. We watched the crows sit on the fence and my father would say, "I wonder which crow is going to pick my bones first?" Once one crow flew down, all the rest would follow. . . . That's what they [the courts] have been doing to my father. He taught me right from wrong. How come they could pick his bones? The judiciary in this state stinks![3]

Although the bitterness and despair of Robert Spring and Hector Berlioz seem just as great despite being separated by 125 years, progress has been made in the area of moral and legal deliberation. Much of this progress has occurred only within the past 5 to 10 years. Two notable examples of this progress are the *Quinlan* and *Saikewicz* cases. In the *Quinlan* case, the family's wish to withdraw life-support systems from an incompetent (but formerly competent) young woman in a chronic vegetative state was held by the New Jersey Supreme Court to be justified on the basis of the constitutional right to privacy. The court wrote:

> No external compelling interest of the state could compel Karen to endure the unendurable, only to vegetate a few measurable months with no realistic possibility of returning to a semblance of cognitive or sapient life. . . . The state's interest [to preserve life] contra weakens and the individual's right to privacy grows as the degree of bodily invasion increases and the prognosis dims. . . . We have concluded that Karen's rights of privacy may be asserted on her behalf by her guardian under the peculiar circumstances here present.[4]

The court went on to state that the Quinlan family and the physician should consult with an ethics committee before deciding to withdraw treatment.

The withholding of life-prolonging treatment in a never competent rather than a formerly competent individual was addressed in the *Saikewicz* decision. It was noted that the substantive rights of an incompetent person were essentially the same as those of a competent individual with respect to the right to decline potentially life-prolonging treatment. For the incompetent individual a substituted judgment could be made. The significant feature of the *Saikewicz* decision was that the Massachusetts Supreme Judicial

Court stated that this decision must be made by a probate judge rather than by the family and physician or even an ethics committee.[5] Many have felt that this decision is unworkable and an embarrassment. The *Spring* case presently serves as a foil for understanding the historical, moral, and legal developments as well as pointing up the present limits of those developments and raising new questions with which we must deal because of modern medicine's constant thrust upon the frontiers of moral and legal deliberation.

Factual Aspects of the *Spring* Case

Earle N. Spring was a 75-year-old retired chemist in 1975. He had been a very intelligent and productive man and had been very busy during his retirement, working on his home, traveling with his wife to Florida in an eighteen-foot trailer each winter, and teaching himself how to repair television sets. However, at about the age of 75 he began to show subtle personality changes, forgetfulness, and an inability to do things that had once been easy. In the late fall of 1977, he developed kidney failure with a rapid deterioration in his renal functioning. Since the cause was not immediately identifiable, he was placed on dialysis treatments while the etiology was pursued, and an effort was made to see if the renal failure was reversible. Over several months it became apparent that the renal failure was not reversible and that the etiology could not be defined. During this period of time there was a rapid deterioration in his overall level of mental functioning such that he became frankly ornery, combative, and disoriented. During 1978, while he received dialysis treatments three times a week, his driving became reckless and his son had to take the car keys away. He began to curse his wife and wander naked outside his house at night. In November 1978, his wife suffered a stroke, and both Mr. Spring and his wife were placed in a nursing home where she recovered slowly to the point of being able to return home while her husband continued to become more disoriented, irritable, and apathetic. In January 1979, Dr. Leroy Shear, Mr. Spring's physician, counseled Mrs. Spring and son Robert, suggesting that they consider having Mr. Spring removed from his dialysis treatments. The family agreed that this should be done. They felt strongly that this was what Earle Spring would have wanted.

In light of the landmark *Saikewicz* decision in the Massachusetts courts, the Spring family filed a petition in the Probate and Family Court for Franklin County, alleging that Earle Spring was mentally incompetent to care for himself and asked that they or some other suitable person be appointed guardian for him so that decisions about the discontinuation of treatment could be handled judiciously. On 25 January 1979, Robert E.

Spring, Earle Spring's son, was appointed guardian, whereupon he requested that the dialysis treatments be stopped because the continued treatment was not in Mr. Spring's best interest considering his age and his physical and mental condition and because Earle Spring would want to decline prolonged medical treatment if the medical circumstances dictated that he would not recover from his illness. The further considerations of pain, discomfort, and suffering related to the dialysis treatments were also cited. At that same hearing, the judge of probate, Sanford Keedy, appointed Mr. Mark Berson, a lawyer practicing in Greenfield, to act as guardian *ad litem* (next friend) to represent the best interest of the ward and additionally "to investigate and report back to the court with reference to the medical procedures that should be accorded the ward." On 9 February 1979, Mr. Berson reported that Mr. Spring was incompetent and unaware that he was being treated by dialysis. He also noted, "I find no expressed wish or desire by Mr. Spring, while he was competent, to withdraw life-supporting systems or treatment, although both his wife and son are of the opinion that if he were competent he would request the withdrawal of such life-supporting systems or treatment." A follow-up hearing was held on 5 April 1979, and Judge Keedy handed down a decree on 15 May 1979 that concluded: "After hearing, it appearing that Earle N. Spring is suffering from an incurable and irreversible illness and that he will not recover from his illness, it is ordered that the temporary guardian refrain from authorizing any further life-prolonging treatment." Mr. Berson, guardian *ad litem,* filed a motion to stay the decree, which was allowed on 23 May 1979. Therefore, Earle Spring continued to receive thrice weekly dialysis treatments against his family's wishes and the attending physician's best judgment. On 2 July 1979, Judge Keedy filed a document entitled "Findings, Rulings and Order for Entry of Judgement," which noted that, if Earle Spring were competent, he would choose not to receive life-prolonging treatment.[6] Although he had not explicitly expressed this wish, he had actively resisted the treatment on occasion by kicking nurses when they attempted to transport him, pulling out needles while he was receiving treatment, and expressing the feeling that he did not want to go for dialysis. Judge Keedy ruled that the decision as to whether to continue or terminate the dialysis treatments should rest with Mr. Spring's attending physician, his wife, and his son. However, the guardian *ad litem* appealed the decision immediately to the appeals court where it was argued on 10 September 1979, with a decision not being rendered until 21 December 1979. This decision affirmed the judgment of the probate court that had been rendered five months earlier. However, the guardian *ad litem* appealed the case further to the Massachusetts Supreme Judicial Court with the case being argued on 10 January 1980. Four days later, this court entered an order that upheld the view that, if Mr. Spring were competent, he would choose not to receive life-prolonging treatment.

However, on a judicial technicality, the full court concluded that the *Spring* case fell within the rule of the *Saikewicz* decision and that the decision to terminate must not be delegated to the attending physician and the family. Therefore, the Supreme Judicial Court remanded the case back to Judge Keedy for entry of a new order that would be technically correct in light of the *Saikewicz* decision. Three days later, on 17 January 1980, another hearing was held in the probate court, and Judge Keedy concluded that, if Mr. Spring were competent, he would choose not to receive life-prolonging treatment and therefore ordered the family "to refrain from authorizing any further life-prolonging treatment, until further order of the court."

The dialysis treatments were then stopped. However, several staff members at the Holyoke Geriatric Center, where Earle Spring resided, objected to the court's decision and allowed an outside nurse and physician who were connected with the Lyndon LaRouche right-to-life movement to examine Earle Spring. They filed affidavits stating that, in their opinion, Mr. Spring was competent enough to express his wishes and that he did desire to live. Judge Keedy denied further consideration of his previous decision on 23 January 1980, but it was immediately appealed to the Supreme Judicial Court that evening with a petition that Judge Keedy's decision be reversed and that new hearings be convened to determine if indeed Earle Spring was competent or incompetent. This appeal was heard by Associate Justice Francis J. Quirico, who entered a temporary order against the withholding of treatment. Thus, Earle Spring was immediately transported to the dialysis center for a treatment that night. In a "Memorandum and Order" dated 4 February 1980, Judge Quirico ordered that because of the gravity and significance of the case, three outside physicians, a gerontologist, a general practitioner, and a psychiatrist examine Earle Spring in order to determine his competency. Judge Quirico wrote in his Memorandum:

> The circumstances of this case warrant, and indeed compel, all concerned to expend whatever time and effort is necessary to determine whether Mr. Spring is currently competent to make or to participate in the making of this most important decision. Surely the question of his competence or incompetence, questioned for the first time so late in these proceedings, is deserving of investigation, particularly in view of the absence of any reasonably current physical or mental examination of Mr. Spring by specialists or experts in the field of the particular question raised. This seems necessary to me notwithstanding the question raised as to the motives or ethics of the two persons whose affidavits were filed by the guardian *ad litem*.[7]

In the ensuing two months, Earle Spring was examined by four separate physicians: a psychiatrist, an internist, a gerontologist who is also a nephrologist, and a neurologist. Their reports were to be filed on 7 April 1980, but Earle Spring died on Easter Sunday morning, 6 April 1980, just before

8:30 A.M. Thus the legal and moral aspects of this prolongation-of-life case remain unresolved, his family remains exhausted and embittered, and the medical community remains confused over its duties and responsibilities. Moreover, the rights of future incompetent patients remain clouded.

Determination of Competency in the Dying Patient

The *Quinlan* and *Saikewicz* decisions presently stand at opposite ends of a judicial spectrum, between which stands the *Spring* case. It therefore stands at the cutting edge of moral, judicial, and scientific interdisciplinary deliberation, and study of this case examines in general the interaction of medicine (science), law, and ethics.

On the basis of reports submitted by the four physicians just prior to Earle Spring's death, it can be stated that Earle Spring was totally incompetent beyond any shadow of doubt. It can be argued that the court's role in seeking to define this was cumbersome, inefficient, costly, and unnecessarily time consuming. It took the court essentially two months to obtain the factual data that could have been obtained within two hours if a more appropriate procedure had been utilized. Because of the purported role of the courts in medical decision making, Earle Spring was maintained on dialysis treatments approximately fourteen months beyond the family's wishes and the better judgment of the attending physician. In the wake of the *Spring* case, it is incumbent upon the medical and judicial community to examine how competency should be determined in cases involving the right to refuse or to withhold treatment through surrogate judgment.

The judicial and medical problems in determining competency are most clearly illustrated in another recent case in the Massachusetts Appeals Court in May 1978. It involved the issue of competency of Mrs. Candura, a 77-year-old woman with diabetes who refused to have her gangrenous right foot and lower leg amputated. Her daughter, recognizing that her mother was becoming senile, obtained guardianship through the probate court and consented to surgery. However, her brothers intervened, arguing that in their opinion Mrs. Candura was competent enough to make a decision to refuse surgery and that her wishes should be honored. The appeals court overturned the probate judge's ruling, finding that Mrs. Candura was competent to refuse surgery despite her periods of irrationality.[8] An examination of that determination of competency by the court reveals some major flaws, if not embarrassment, for both the judiciary and the medical community.

Historically, the declaration of incompetency was considered to be a human act, protecting wrongdoers from punishment for acts for which they were not responsible and protecting the mentally ill from carrying out acts that might be detrimental to their well-being. Labeling someone legally incompetent, however, can be detrimental to the personal dignity and integrity of that individual. This has certainly been recognized in the mental health law field. For the dying individual it may represent an infringement on his or her right to a human and dignified death, free of suffering. How is competency to be determined then?

In U.S. common law, an individual is considered competent until determined otherwise by a judicial process. Historically, the assessment of mental competency to stand trial in a criminal proceeding has been at the forefront of law and medicine. This area has been a stimulus for the development of the whole subspecialty of forensic psychiatry. During the early decades of this century, psychiatrists began to assume an active role in aiding the courts by serving as consultants and expert witnesses in criminal proceedings as they were deemed to have the greatest degree of expertise in assessing someone's mental state.

The role of the psychiatrist first became formalized in statutory law in the United States in 1921 with the passage of the Briggs Law of Massachusetts. This law, named after its chief psychiatric advocate, Dr. Vernon Briggs, was the first statute to require an automatic pretrial psychiatric evaluation of certain classes of criminal defendants. The primary intention was to eliminate the battle of the experts that at the turn of the century had become a spectacle of public ridicule and professional disenchantment in criminal proceedings involving psychiatric factors. It was hoped that, by having two psychiatrists designated by the state's department of mental health, examinations would be carried out in an impartial manner and that the findings and opinions would be accepted by the respective parties.[9]

On the surface, the association of psychiatry and the law would appear to be a natural one as the legal connotations of the phrase *mental competence* would appear to be similar to the psychiatric connotations. However, the semantic implications are not actually identical, and much confusion has, therefore, frequently arisen. In the legal sense, when applied to the criminal it refers to whether the individual demonstrates a degree of mental soundness sufficient to stand trial. Identical criteria are used without regard to the offense. When applied to civil proceedings, civil responsibility is adjudged in terms of the degree of mental competence appropriate to the particular act in question—for example, writing a will, making a contract, refusing medical treatment, or obtaining guardianship. Traditionally, the criteria for determining competency have varied with the act. For example, the criteria for determining whether an adult is competent to be the guardian of a child are more demanding than the criteria for determining whether an individual can dispose of his own possessions by making a will. Freedman has noted how difficult this assessment can become in the frequent situation in which allegations are made "of senile deterioration, chronic psychosis, mental retardation, or incapacitating chronic alcoholism."[10] In these disease states, the psychiatrist has particular difficulty in answering two fundamental questions: (1) whether the patient understands the nature of the specific act in question and (2) whether the patient is aware of his duties and obligations in connection with the act.

The Determination of Competency in *Lane* v. *Candura*

Analysis of the psychiatric testimony in the *Candura* case provides a poignant example of the potential confusion that can arise in determinations of competency in brain-damaged individuals. In turn, it highlights the limitations of contemporary forensic psychiatry, especially when carried out from a psychoanalytic perspective.

The first psychiatrist, a Dr. Kelley, stated in his testimony that in his opinion Mrs. Candura was incompetent to make a rational choice. His view was based upon three inferences derived from his examination of her: "(1) his inference, from her unwillingness to discuss the problem with him, that she was unable to face up to the problem or to understand that her refusal constituted a choice; (2) his characterization of [her] . . . as suicidal; and (3) a possibility . . . that her mind might be impaired by toxicity caused by the gangrenous condition."[11] (It is interesting to note that the court of appeals particularly refuted this last inference as it could not be supported by clinical evidence.) Relying heavily upon Dr. Kelley's opinion, the probate judge concluded that:

> Without necessarily finding that the ward is mentally ill for all purposes, . . . she is incapable of making a rational and competent choice to undergo or reject the proposed surgery to her right leg. To this extent, at least, her behavior is irrational. . . . The court finds that her confused mental condition resulting from her underlying senility and depression warrants the exercise of the jurisdiction of the court.[12]

The second psychiatrist, a Dr. Zeckel, felt that the patient was competent despite her forgetfulness and disorientation in time. His examination of her revealed the following reasons for her refusal of the surgery: (1) the fact that she had been unhappy since the death of her husband, (2) the fact that she did not wish to be a burden to her children, (3) the fact that she did not believe the operation would help her since the first bypass operation had failed to arrest the advance of gangrene, and (4) the fact that she did not wish to live as an invalid or in a nursing home. In testimony, she also indicated that although she ultimately desired to get well, she felt that this was no longer a realistic possibility. Thus, she did not fear death but welcomed it.

The subjectivity and limitations of the presumed expert psychiatric testimony are strikingly highlighted in the court's investigation into the conflicting testimony. When asked to explain why his opinion differed from that of Dr. Kelley, Dr. Zeckel answered, "I think it is just a personal philosophy type of thing where I believe the person(s) themselves ought to be given the benefit of the doubt as to what they want to do with their lives, whereas, Dr. Kelley, I guess, is more protective. I can't really speak for him, but his general philosophy is different from mine."[13] Likewise, a portion of Dr. Kelley's testimony is of interest: "You know, it comes down really to a philosophical [difference]; I hope there is no psychiatric argument in this case. It's the right of a patient to decide they want to die, and I spend all my life trying to keep people alive, so I take quite a different view."[14]

Behavioral Neurology

We have seen in the preceding account that, based on the facts obtained in the case, two psychiatrists and two courts reached different conclusions regarding Mrs. Candura's civil competency to refuse medical treatment. Although several explanations for this might be offered, the most important one is that the methods of contemporary forensic psychiatry, especially when carried out from a psychoanalytic perspective, are inadequate for collecting clinical data concerning mental functioning in organically impaired patients. Second, the methods of contemporary forensic psychiatry are inadequate for organizing the data into a coherent framework in order to subsequently evaluate the facts. The application of recent advances in the clinical specialty of behavioral neurology would seem to offer a means for improving the collection of data and organizing it into a conceptual framework based upon clinical phenomena as they arise in nature. In addition, behavioral neurology appears to uniquely complement the Anglo-Saxon common-law notion of competency appropriate to the act in question.

Behavioral neurology is that branch of clinical neurology concerned with the analysis of human behavior and intellectual functioning through the discipline of human neuropsychology. Human neuropsychology, in turn, is based upon a systematic analysis of disturbances of behavior following alterations of normal brain activity by disease, damage, or experimental modification. In addition to systematic observations of abnormal behavior, neuropsychology offers the possibility of determining cerebral localization of mental functioning and, thus, of correlating the components of intellectual functioning to specific areas of the brain.

Behavioral neurology presently attempts to analyze human behavior and intellectual functioning in terms of seven parameters, or states:

1. Arousal and attention,
2. Emotion (mood and affect),
3. Language,
4. Cognition,
5. Memory,
6. Praxic and constructional ability, and
7. Thought content.

Behavioral neurology utilizes a detailed but flexible battery of bedside or complex neuropsychological tests to examine each of these parameters of mental functioning in far greater detail than present-day forensic psychiatry. More important, however, it then attempts to interpret the findings in light of its constantly evolving theories of the neurobiologic basis of human mental activity and behavior.

In behavioral neurology, a constant tension exists between attempts to localize mental functioning and behavior to specific areas of the brain and attempts to understand how these specific functions and behaviors with their localization are integrated and interdependent. For example, behavioral neurology, through its study of disease states, has found that disorders of memory can exist relatively independent of language, praxis, emotion, and cognition. In turn, disorders of memory can exist relatively independent of each other. In Korsakoff's syndrome, for example, immediate recall and remote memory can be intact while recent recall and the ability to learn are profoundly impaired. A further example is that the ability to reason and to abstract can be preserved in the face of disorders of memory and language.[15]

Mrs. Candura's case of senile dementia illustrates this. Although she was found to be "irascible, . . . disoriented to time, . . . forgetful and to have a wandering train of thought," the court of appeals noted that "she exhibited a high degree of awareness and acuity when responding to questions concerning the proposed surgery." Thus, the court of appeals concluded (and behavioral neurology would argue, justifiably so) that "senile symptoms, in the abstract, may of course justify a finding of incompetence, but the inquiry must be more particular." Behavioral neurology, thus, offers a powerful tool for aiding that particular inquiry in brain-damaged and dementing patients by improving the collection of neuropsychological data and fitting it into a coherent framework of mental functioning based on particular disease states.

An Ethical and Legal Implication of Behavioral Neurology

The import of behavioral neurology does not lie just with its methodology and theoretical groundwork. Potentially, it also has significant implications for our understanding of human nature. In turn, it may well have implications for general theories of ethics and for particular issues in medical ethics with respect to dementing and brain-damaged persons such as retarded persons and stroke victims.

Specifically, behavioral neurology has implications for our understanding of the important legal and ethical notion of personhood. Historically, Anglo-Saxon philosophy and law have given man's cognitive capacities ascendancy over his emotional capacities. In this tradition his emotions have been considered untrustworthy and often animal-like. Reason and language are what differentiates man from other organisms on the phylogenetic scale. In this context it is of interest to review John Locke's definition of a person since Locke's philosophy stands as one of the cornerstones of our Anglo-Saxon image of man:

> A thinking intelligent being, that has reason and reflection, and can consider itself as itself, the same thinking thing, in different times and places: which it does only by that consciousness which is inseparable from thinking, and seems to me essential to it.[16]

However, modern-day neurobiology's understanding of the anatomy, physiology, and pathology of the emotions strongly indicates that human emotional capacities are unique and not to be equated necessarily with animal instincts. Human emotions and will should be considered an integral part of what it means to be human and to have personhood. The evolutionary ascendancy of human emotional development may be just as much related to the phylogenetic expansion of his telencephalon as his cognitive, linguistic, and mnemic ascendancy. Recent clinical studies suggest that the right hemisphere and neocortex play a unique role in integrating the human emotional state as organized in frontolimbic systems.

Thus, specific syndromes exist in clinical neurology that affect an individual's rationality more than his emotional capacities and integrity. Senile dementia prior to the onset of apathy, as exemplified by Mrs. Candura, is one such clinical pathologic entity. In such individuals, should not our ethical and legal deliberations give as much recognition of and protection to this aspect of one's personhood? In the compromised individual, should not the avoidance of pain and prolonged suffering, organized and controlled by frontolimbic and perhaps right-hemispheric functioning, be given as much recognition and protection as rationality? Pascal once wrote: "The heart has its reasons, which reason does not know." Translated into a contemporary neuropsychological framework, Pascal's wisdom might read: The frontolimbic system has its justifying value that reason can never know.

Notes

1. John Ferriar, *Medical Histories and Reflections* (Philadelphia: Thomas Dobson, 1816).

2. Sir Robert Wilson, "A Medico-Literary Causerie," *Practitioner* 56 (1896):631–635.

3. Jean Caldwell, "Earle Spring Remembered through the Eyes of His Son," *Boston Globe*, 9 April 1980. (Reprinted courtesy of the *Boston Globe*).

4. *In the Matter of Karen Quinlan*, Supreme Court of New Jersey, 70 N.J. 10, 355 A.2d 647, 31 March 1976.

5. Superintendent of Belchertown State School v. Saikewicz, 373 Mass. 728 (1977), Mass. Adv. Sch. (1977) 2461.

6. *In the Matter of Earle A. Spring. Memorandum and Order*, Supreme Judicial Court for Suffolk County, no. 80-37 Civil. 4 February 1980.

7. Ibid.

8. Grace R. Lane v. Rosari Candura, Massachusetts Appeals Court (Middlesex) 78-417, 26 May 1978.

9. William J. Curran and E. Donald Shapiro, *Law, Medicine and Forensic Science* (Boston: Little, Brown & Company, 1970).

10. Lawrence Z. Freedman, "Forensic Psychiatry," in *Comprehensive Textbook of Psychiatry*, edited by Alfred M. Freedman and Harold I. Kaplan (Baltimore: Williams and Wilkins Company, 1967).

11. Lane v. Candura.

12. Ibid.

13. Ibid.

14. Ibid.

15. Henry Hecaen and Martin L. Albert, *Human Neuropsychology* (New York: John Wiley & Sons, 1978).

16. John Locke, "An Essay Concerning Human Understanding," Book II. vol. 27, p. 9.

20 Life Preservation and the Incompetent

Richard A. McCormick

Chapter 19 reviewed some of the significant cases regarding this issue of life prolongation and competency—*Quinlan, Saikewicz,* and *Spring*. A more recent case, one probably familiar to readers, is the *Brother Fox* case in New York that applies now because it has been a decision of the appellate division of the Supreme Court. It applies throughout New York State, and it is dreadful in my judgment. I take it that I share Dr. Hardy's judgment that we are being overwhelmed by procedures and not being allowed to die in a human way.

The *Brother Fox* case was basically and simply this: An 83-year-old brother of the Society of Mary was having hernia surgery on 2 October 1979. He suffered a heart arrest of between two and five minutes' duration. There was massive anoxia of the brain stem, and he ended up in a Karen Ann Quinlan condition, essentially in a vegetative state with no hope of recovery. His superior, Father Philip Eichner, attempted to get him removed from the respirator. Nassau Hospital refused, Father Eichner appealed, and was given relief. However, the district attorney decided he would appeal the decision because the issues were so important. In the meantime, Brother Fox died on 24 January 1980 on the respirator.

In March 1980, the appellate division of the Supreme Court of New York gave its decision, essentially granting relief. The decision established for us the procedures to be used in the future—namely, that every comatose, terminally ill patient must go to the court in order to be removed from a respirator. First, three doctors must examine the patient to make sure that the illness is terminal, that there is no hope for recovery. Then the district attorney and the attorney general must be informed, at which point they have their doctors, if they so wish, view the patient. Then all these things are heard in court.

If you stand back from cases like this—*Saikewicz* and now *Fox*—I think you see a larger pattern developing, and I think it is ominous in its drift. First of all, management decisions in medicine, in my judgment, if they are to remain truly human and promotive of our individual best interests, ought to be controlled within the family/patient/physician relationship. They are a glove to fit an individual hand. They involve far more than the medical; they involve aspirations, ideals, religious beliefs, a whole host of personal things. Therefore, they are very personal judgments. What is happening in our time, I believe, is that three very impersonal factors are increasingly dictating our treatment of the dying: (1) high technology, (2) cost-containment factors, and (3) public entities such as the law and the courts. Taken together, those constitute three impersonal intrusions into what is essentially a personal decision. Therefore, I see that as the major threat in the type of history that Dr. Hardy was proposing to us in the last chapter. Not that there should not be recourse to the courts; of course there must be. We can make abusive decisions, but we do not start with a resounding vote of no confidence in families and physicians working together. We establish certain presumptions.

Now let me turn, therefore, to another incompetent group: babies. After all, we must make proxy or surrogate decisions here, substituted judgments; otherwise all ill infants are simply left on machinery. Somebody has to say yes or no, and the important thing is, it seems to me, if we are going to control these impersonal factors that I mentioned, that we as a thinking community develop precise and disciplined convictions in these areas. Otherwise, these impersonal forces simply will continue in their own way with a new (pardon my manifest bias here) priesthood of the judiciary taking over what are essentially very individual personal decisions. Therefore, it is important for us to think clearly about this. Let me begin with one simple case of an incompetent in order to force our attention concretely on the type of thing I would like to discuss.

Baby-boy Baxter was born with a severe heart abnormality. Immediate corrective cardiac surgery was required. The surgery revealed an opening in the septum separating the ventricle and then malformation of the mitral valve. Surgical reconstruction dealt with the immediate problem. A week after surgery, cardiac insufficiency was clearly going to necessitate further surgery. The surgeons explained in detail to the parents the seriousness of the problem and the bleak prognosis. They predicted one chance in ten of long-term survival, but only with a very long series of heart operations—doctors projected as many as twenty. The repeated surgery and recovery would be physically painful for the infant—again with low prognosis. The parents decided that the extreme burden was not morally necessary and, with the concurrence of the surgeon, rejected further surgery. The child died the following week. I imagine we all have our own instinctive reactions to that procedure. Was it right or wrong? I think it was proper within the parameters of what is Christian and human to say that we have gone far enough with what we have. We can save a life perhaps here but at such an enormous expense that one can doubt that it really renders a benefit to the one who is saved. That is the type of case I raise for discussion.

We have all kinds of cases like this—for example, spina bifida with meningomyelocele, necrotizing enterocolitis, premature babies, of course, whom the more you treat sometimes the more you risk injuring, and so on. When can we call a halt to these things, as concerned parents, concerned physicians, and concerned public? In order to get a focus on how we think ethically in these areas, both of adult competence and infant incompetence, I would like first to define what I consider to be the shape of a bioethical problem. Basically, most bioethical problems under analysis, when stripped of their technical materials, will reveal that a basic human value, around which we have developed a policy of protection and promotion and a vocabulary to communicate that policy, has come into new circumstances, usually technological, forcing us to ask new questions. Take for example another area—childbearing, parenting, and the notion of the family. Certainly those are very basic values, and we have protected them with certain policies. We have developed the language to communicate that, not definitive and unchanging but something that we feel comfortable with, to promote our overall well-being. Suddenly we have come into new circumstances, technological in nature, like in vitro fertilization with embryo transfer, the so-called test-tube baby. That is not just a matter of doing something a little artificial. Some very basic values here are at stake: the very notions of parenting and of sexual love, the identity of the child, and so on. The same thing has happened in the area of preservation of life. It is no longer a matter, in my judgment, of merely preserving a life. We have the technology to do that almost indefinitely, almost with anyone. The question now is: At what point is this preservation a genuine benefit in the best interest of this patient? At what point does preservation cease to be a benefit to this patient? This is a different type of question because of the enormous technology available to us. In attempting to face this question, our tradition (I say our, speaking as a Catholic theologian, but this is shared by a broader Judaeo-Christian tradition) has attempted to walk a middle path between two extremes. On the one hand is the extreme we will call medical moral utopianism or vitalism—that is, you keep life going at any cost, with any means available, no matter what the condition of the patient, because when life is over, everything is over. On the other hand is the extreme of medical moral pessimism—that is, when life becomes difficult, dysfunctional, tedious, and so forth, you kill. Therefore, it is just the opposite side of that coin. Between those extremes, I judge that our tradition has attempted to walk a middle path with a single value judgment and then protect that with a vocabulary. That value judgment would be roughly that human life is the source of all other experiences and values and is a basic good, the most urgent basic good, but not an absolute one. Or, if you turn the coin over, you could say that death is, in the ordinary run of things, an evil to be avoided but that it is not an unconditional or absolute evil since there is a time to die. That basic value judgment is at the heart of our traditional teaching.

Therefore, let me move in two steps. I first describe the traditional teaching and vocabulary and, second, move into the contemporary period where I judge in our time we have new problems and therefore must reformulate and regather our forces to see what we are going to do about some of the problems that have been raised in this book.

In the traditional approach, because life is a basic good but not an absolute one, there is a limit on the means we must use to preserve that life. That limit has been stated in terms of ordinary and extraordinary means. Pius XII used this language, and it has been in our Catholic tradition for decades. It was adopted officially by the American Medical Association in 1973 by the House of Delegates and so on, so it is very common language. Basic to it, though, is the value judgment that life is a basic good but not an absolute one. If that is true, not all means must be used since, if we had to use every means available to us, we would force our attention and our energies, all our resources, on a good in such a way that we would distort its real meaning. We would absolutize that good if we had to use all means.Therefore, as Pius XII repeatedly stated, there is a limit on what we must do. At some point, therefore, excessive concern for the temporal is neglect of the eternal. That limitation has always been stated in terms of ordinary and extraordinary means.

Very quickly, these terms have become badly misunderstood; they were badly abused in the *Quinlan* case, in the now famous *Johns Hopkins* case, and in contemporary discourse in general. Most people use the term *ordinary* to refer, first, to what doctors ordinarily do when faced with a certain crisis. Second, if a very highly experimental procedure is involved, doctors do not ordinarily do it and it is therefore extraordinary. In other words, the reference is to a single aspect—what doctors customarily do when faced with certain crises. That is exactly what the term does *not* mean, at least as it should be used ethically. When you use the terms *ordinary* and *extraordinary* in the ethical or the moral sense, you have reference to a whole host of other circumstances of the patient, not just what doctors do.

Therefore, the definition of an ordinary means in the moral sense has been the following: any means, medicine, surgical intervention, therapy, and so on, that can be obtained and used without excessive inconvenience (no grave inconvenience) and that offers reasonable hope of benefit to the patient. Now notice how general the phrase *excessive inconvenience* is. When we ask what excessive inconvenience is, it becomes very relative to all of us as patients. Consider the phrase *reasonable hope of benefit,* and we need to ask what a reasonable hope of benefit is. Benefit is quite different, for example, for a two-year-old child than it would be for a 92-year-old, obese, third-degree cardiac patient. Thus, these are very relative terms but were the two aspects that had important function. If something could be had and used without real difficulty, or great inconvenience, and if it offered me some reasonable hope of benefit, then the means was said to be ordinary and I had to use it. Otherwise I was committing suicide. On the contrary, if the means (surgery, medicine, therapy, whatever) involved either grave inconvenience or, even if it did not involve such an inconvenience, was of no real benefit to me, then it was said to be extraordinary, and I need not use it. It was morally dispensable. Among the inconveniences that were always mentioned was, for example, pain in preanesthetic days when the only anesthesia would have been a few drinks of whatever was available. Expense is another example of inconvenience: In some medical interventions a family would be rendered destitute, its savings gone. The other members of the family would be deprived of education, and that was considered a cost I need not pay as the father of the family. I need not save my life at that cost to my dependents. So pain, cost, danger, and the experimental character of the procedure could create grave inconvenience.

We used to do routine appendectomies on people going to certain missionary countries before they left, because when they got there such an operation might not be possible. What has happened in our time is that of those two elements—some reasonable hope of benefit and no grave inconvenience—grave inconvenience is increasingly not functioning in life-preservation decisions. It does not mean it is out of consideration, but with modern medicine's good pain management, it is usually not a consideration in many of these judgments. Exceptions exist of course, but this is the trend. With third-party carriers, expense is often not a determinant in life-and-death considerations. Eventually we are going to reach an end to this determinant because, for example, in the neonatal intensive-care unit at Georgetown, a baby's bill there for four months will be $100,000. Someone must pay it; but it does not function in most of these life-and-death decisions.

Thus, the attention is increasingly focused on a single element—namely, the benefit to the patient. We can save this patient, but in what condition are we going to leave him or her? We can intervene with this baby or with this very elderly incompetent person, but what genuine benefit are we going to do by extending the life of Joseph Saikewicz for six months or three months or Brother Fox (in a chronic vegetative state) by half a year? That type of decision is bothersome because it forces us not to simply give an answer. The least interesting thing about ethics is the answer. Rather it is how you arrived at the answer because that controls other decisions. Thus, I judge that this terminology—ordinary and extraordinary—has reached the limit of its usefulness because it now hides the judgments that we are making. Take the case at Johns Hopkins that I mentioned. Here was a baby born with Down's syndrome and with duodenal atresia—a blockage of the duodenum preventing the ingestion of food. That condition is easily correctable by surgery. The parents refused, with the advice of their priest, to allow interventive surgery, saying that the means were extraordinary. The child was put in a bassinet with a sign on it, "No Food by Mouth," and of course if you fed it in that condition you would choke it—kill it. If you did not feed it, you would starve it. Its death fifteen days later was justified on the basis that the means were extraordinary.

Notice what is happening there. The terms being used are hiding the real judgment. The real judgment was that the condition of this child was extraordinary. The kind of life you will give it by operating and keeping it alive was judged to be nonworthy. I think that decision was wrong in all respects. Nonetheless, it shows the weakness of the terminology. The same thing happened in the *Quinlan* case. Notice how Bishop Lawrence Casey, in his amicus brief, put this when talking about Karen Ann Quinlan. He said that Karen Ann had no reasonable hope of recovery from her comatose state by the use of any available medical procedures. The continuance of mechanical supportive measures to sustain continuation of her body functions and her life constituted extraordinary means of treatment. This really involves two sentences: (1) We have an extraordinary means, and (2) It is extraordinary because all we can do is keep her in the condition she is in, not give her anything but a chronic vegetative state. What benefit are we visiting on the patient by intervening? Obviously, these are quality-of-life judgments. I hate to use that language, and I am going to forfeit its use in the future because of the type of emotional ripple it creates. You know we had a destructive path in the Third Reich, leading from the nonrehabilitable sick to the inclusion of everyone in the non-German category.

Therefore, I think our challenge is to find criteria in our time that will define what is a genuine benefit to the patient. Let us abandon the element that is so often present—namely, maintenance of simple body functions like respirating and circulating. I assume we do not preserve life just to have that, to have a chronic vegetative state. What minimal conditions beyond that should we be looking for so we can say we are providing a benefit by intervening? That is the question for our time. The words ordinary/extraordinary hide the particularity of what we are looking for. In pursuit of this question over a period of years, I have been led in a single direction and by, above all, several considerations. Let me just give you an instance of them.

Take Paul Ramsey's book *The Patient as Person.*[1] Paul Ramsey has provided insight to all of us in the care of the dying. He points out that our basic obligations with regard to those who are desperately ill, first, is to cure when we can and, second, when we cannot cure, to care, to company with. This is what hospices are all about. There comes a time, however, in the dying process, at least in some individual cases, at which we cannot care and company any more. He mentions two conditions. First is the condition of profound intractable pain—pain that excludes companying because profound unmanageable pain in the dying process isolates the individual into his or her own experience in such a way that others are excluded. The second condition is profound and irretrievable unconsciousness in the dying process. Ramsey says that when that happens it makes little difference how death achieves its victory, which is another question, of course. At any rate, we cannot company with those in coma. What I find very useful there is that I think he put his finger on the basic two conditions beyond simple corporal circulation and ventilation that we are looking for: (1) absence of severe, protracted unmanageable pain and (2) some minimal degree of human interrelating.

Now another indicium that has led me in this direction is Sissela Bok's article in the *New England Journal of Medicine* on the living will. She used the opportunity to criticize several versions of the living will that were then available and suggested her own. I daresay I would write mine the same way. I hope many of you will share in these feelings:

> I wish to live a full and long life, but not at all costs. If my death is near and cannot be avoided and if I have lost the ability to interact with others, [one condition], or if my suffering is intense and irreversible [the other condition], I do not want to have my life prolonged. I would then ask not to be subjected to surgery or resuscitation nor would I then wish to have life support from mechanical ventilators, intensive care services, or other life-prolonging procedures including the administration of antibiotics and blood products. I would wish rather to have care which gives comfort and support, which facilitates my interaction with others to the extent that this is possible, and which brings peace.[2]

So the very same things that we noticed in Ramsey have been mentioned here: loss of the ability to interact with others with no chance of regaining it or intense and irreversible suffering. I suggest, therefore, that in many decisions involving management and even life and death with regard to the incompetent (including babies), these are the criteria that will often function. If all we can offer a young baby, for example, by intervening with resuscitative devices or surgery and so on, is continuance in a state that has no, absolutely no, prognosis for any interaction with other people, then is this a wise intervention?

On the one hand, I think that in terms of my own tradition (the way I read it), we would be better off allowing that baby peacefully to pass on to a future life; we are offering no benefit to the patient, at least as I judge it. On the other hand, if all we can offer the child is life support with repeated intrusions that are extremely painful, combined with very low prognosis, then is there not a point when we have gone far enough? These are very simple general statements. They do not solve our problems, but they do align us in the direction, I think, of many of the decisions that must be made with regard to the incompetent, both the adult and the infant.

The difficulty here is, of course, that we are dealing in, as Dr. Hardy said in chapter 19, substituted judgments; we must make the decision for someone else. That is fairly clear when family history has been close and when individual aspirations of family members are known. They have had a life, they have had a chance to etch their biography upon those with whom they live. We can help; we can make these substituted judgments with a certain security.

When dealing with babies, however, there is no knowledge of the individual, there has been no history, the decision is utterly final. Therefore, we must favor life whenever there is a reasonable doubt. It seems to me that is one of the exception stoppers, one of the cautions, of someone who is truly prolife in the broadest and best sense. That must be our conclusion. These decisions must be made (very often) at a time when prognosis of any kind of future is very uncertain. My physician friends insist on that. Furthermore, whatever decision we make in these cases is generalizable to all babies. This explains the enormity of the responsibility upon us in trying to make substituted judgments. However, when all those cautions are built in (benefit of the doubt to life and so on), I believe it is true to say that there are cases of children that we could save but whose saving would not represent a benefit to them in any sense that I understand as being Christian. I would put attempts to save in some cases in the category of what I call medical moral optimism or vitalism—that is, simply saving the life in order that there be ventilation and circulation and that our hands be clean.

Now let me conclude by recalling attention to the Quinlan case. This case referred, at one key point, to what most of us would do were we in Karen's position. That is a substituted judgment, but behind that, very clearly, is an appeal to the reasonable person—what we as reasonable persons would do were we in such a condition. I think, from what I see in the drift of the courts and decisions now available to us, that there will be increasing reference to substituted judgments based on the assessment of reasonable people. We thus have the enormous importance of making ourselves, by study and discussion like this, knowledgeable, reasonable people.

Notes

1. Ramsey, Paul, *The Patient as Person* (New Haven: Yale Unversity Press, 1975).

2. Bok, Sissela, "Personal Directions for Care at the End of Life," *New England Journal of Medicine* 295 (1976):369. Reprinted by permission.

About the Contributors

David A. Begelman is a clinical psychologist associated with the Psychiatric Clinic at Charlotte Hungerford Hospital in Torrington, Connecticut. He has a background in art and philosophy in addition to his doctorate in psychology. Dr. Begelman has taught at Yale University and at Kirkland College, and is the author of numerous articles in psychology, behaviorism, and ethics. He also serves as a consulting editor for behavioral psychology journals.

John D. Biggers is professor of physiology at Harvard Medical School and a member of its Laboratory of Human Reproduction and Reproductive Biology. He is former president of the Society for the Study of Reproduction and former editor-in-chief of the *Journal of Biology of Reproduction*. Professor Biggers is the author of many scientific articles dealing with reproductive physiology, tissue culture, and growth and differentiation. He served as a consultant to the Ethics Advisory Board of the Department of Health, Education, and Welfare (now the Department of Health and Human Services).

Howard Brody received the M.D. and Ph.D. in philosophy from Michigan State University in East Lansing, and completed his residency training in family practice at the University of Virginia. Dr. Brody is currently assistant professor of family practice and philosophy and assistant coordinator of the Medical Humanities Program at Michigan State. He is author of *Ethical Decisions in Medicine* (2d ed., 1981) and *Placebos and the Philosophy of Medicine* (1980). He formerly served as a council member of the Society for Health and Human Values.

Barton Childs graduated from Williams College and received the M.D. at The Johns Hopkins University School of Medicine. His internship and residency were in pediatrics at Johns Hopkins Hospital. Dr. Childs is now professor of pediatrics at The Johns Hopkins University School of Medicine. He is a recipient of the E. Mead Johnson Award and the William Allan Award of the American Society of Human Genetics.

Charles E. Curran is currently professor of moral theology at the Catholic University of America in Washington, D.C. Father Curran, a Roman Catholic priest of the diocese of Rochester, New York, received doctorate degrees in theology and moral theology from the Gregorian University (Rome) and from the Academia Alfonsiana (Rome) before beginning his teaching career at St. Bernard's Seminary in Rochester, New York (1961–1965). He is a past president of both the Catholic Theological Society of America and the American Society of Christian Ethics. In 1972, he was named the first recipient of the John Courtney Murray Award of the Catholic Theological Society of America for distinguished achievement in theology. He is the author of more than ten books and has coauthored and edited other volumes. His latest works include *Themes in Fundamental Moral Theology, Issues in Sexual and Medical Ethics,* and *Transition and Tradition in Moral Theology.*

Burr Eichelman is chief of psychiatry service at the William S. Middleton Memorial Veterans Hospital. He is also associate professor of psychiatry and director of the Laboratory of Behavioral Neurochemistry at the Waisman Center of the University of Wisconsin. He is a researcher in the biological bases of aggressive behavior and a writer on ethical issues in aggression research and treatment.

John C. Fletcher serves as assistant for bioethics in the Clinical Center of the National Institutes of Health (NIH) in Bethesda, Maryland. He is an Episcopal minister and has served in parishes in Alabama and Virginia. Based on a study of the ethics of medical research at the Clinical Center of the NIH, he received the Ph.D. in the field of Christian ethics from Union Theological Seminary in 1969. His research interests are in the ethics of applied human genetics, the costs of terminal illness, and informed consent.

Paul Millard Hardy, is assistant professor of neurology at the Tufts-New England Medical Center and Tufts University School of Medicine. He also holds a joint appointment at the Harvard Medical School as a clinical instructor in neurology. Dr. Hardy is a member of the Ethics Committee of the American Academy of Neurology and serves as chairman of the Ethics Advisory Committee for the Massachusetts Huntington's Disease Center Without Walls. He is a former Joseph P. Kennedy, Jr., Fellow in medical ethics at Harvard University. His main clinical and research activities involve the behavioral and neuropsychiatric disorders of the mentally retarded.

Leo J. Hetling received the B.C.E. from Manhattan College and the M.S. and Ph.D. in environmental engineering from Rensselaer Polytechnic Institute. He has held a number of different positions in the New York State Department of Health and has also served in the Federal Water Quality Administration. Dr. Hetling was director of research for the New York State Department of Environmental Conservation and is currently director of the Division of Environmental Health of the New York State Department of Health. He is the author and coauthor of many professional publications in the field of water-quality control.

Ruth Hubbard is professor of biology at Harvard University, where she teaches courses dealing with interactions between science and society. Her laboratory research has emphasized the biochemistry and photochemistry of vision. During the past ten years, she has concentrated her research, writing, and lecturing on the politics of health care and health issues, especially issues of relevance to women. She is coeditor of two collections: *Women Look at Biology Looking at Women* and *Genes and Gender II: Pitfalls in Research on Sex and Gender*. She is currently trying to formulate a contextual biology that takes account of the multiple and complex interactions between ways of living and bodily functions.

John Ladd is professor of philosophy and chairman of the Program in Biomedical Ethics at Brown University. He is the author of *The Structure of a Moral Code* (1957), translator and editor of Kant's *Metaphysical Elements of Justice* (1965), and editor of *Ethical Relativism* (1973) and *Ethical Issues Relating to Life and Death* (1979). Professor Ladd is past president of the American Society for Political and Legal Philosophy and past chairman of the Committee on Philosophy and Medicine of the American Philosophical Association.

Daniel C. Maguire received the S.T.D. degree in theology from the Gregorian University in Rome. He has taught at Catholic University in Washington, D.C., and is now professor in the Department of Theology at Marquette University. He is also the president of the North American Society of Christian Ethics (1981–1982). He is the author of *Death by Choice, The Moral Choice, A New American Justice: Ending the White Male Monopolies,* and many other books and articles in the area of ethics.

Richard A. McCormick received the doctorate in theology from the Gregorian University. Father McCormick is presently Rose F. Kennedy Professor of Christian Ethics at the Kennedy Institute of Ethics and research associate at the Woodstock Theological Center. He is the author of numerous books and articles on Christian theology, morality, and medical ethics. He has served on the Ethics Advisory Board of the Department of Health, Education, and Welfare and has been a Hastings Center Fellow.

Edmond A. Murphy, professor of medicine at The Johns Hopkins University, holds the degrees of M.D. (Queen's University, Belfast, 1952) and Sc.D. (The Johns Hopkins University, 1964). His training has been in internal medicine (at London) and in medical genetics (at The Johns Hopkins University). His more than 200 publications deal with vascular disease, theory of medical genetics, and the fundamental theory of medicine. He is author of *Principles of Genetic Counseling* (1975), *The Logic of Medicine* (1976), *Probability in Medicine* (1979), and *Skepsis, Dogma and Belief* (1981).

Richard A. Perlmutter received the B.A. from the State University of New York at Buffalo with a major in philosophy, attended medical school at the University of Pennsylvania, and interned at Hartford Hospital in Connecticut. He completed the residency in psychiatry at the University of Rochester where he is now clinical director of the Psychiatric Emergency Department. In addition, he serves as a consultant to two university health services.

Warren T. Reich has been senior research scholar in the Center for Bioethics, the Kennedy Institute of Ethics, at Georgetown University since 1971. In addition, since 1977 he has been professor of bioethics and director of the Division of Health and Humanities in Georgetown's Department of Community and Family Medicine. He is editor-in-chief of the *Encyclopedia of Bioethics,* published in 1978—the first encyclopedia in its field. His degree, the S.T.D. in theological ethics, was earned at Rome's Gregorian University.

Mark Sagoff is a research associate at the Center for Philosophy and Public Policy, University of Maryland. After receiving the Ph.D. at the University of Rochester, he taught philosophy at Princeton, the University of Pennsylvania, the University of Wisconsin (Madison), and Cornell University. His work at present is supported by a fellowship from the National Science Foundation and the National Endowment for the Humanities.

Roger C. Sider is associate professor of psychiatry at the University of Rochester School of Medicine and Dentistry where he is also director of Adult Ambulatory Services, director of Undergraduate Education, and co-director of the Program on Psychiatric Ethics. His publications have been in the fields of clinical psychiatry, medical education, and the interface of psychiatry and religion.

Andrew A. Sorensen is associate professor and associate chairman of the Department of Preventive, Family and Rehabilitation Medicine at the University of Rochester School of Medicine and Dentistry. He is an ordained minister in the United Presbyterian Church in the United States, and received the Ph.D. in medical sociology from Yale University. In addition to his interests in medical sociology, he has written in the fields of addiction, health services research, and medical ethics.

Dorothy H. Summers started her nursing career in 1944 and trained at the

Royal Northern Hospital, London, where she held posts as a staff nurse and ward sister. After obtaining the nurse-tutor diploma of the University of London, she taught in the School of Nursing at South London Hospital for Women and Children for nine years and subsequently became a senior tutor at Mayday Hospital, Croydon, where she worked for ten years. During this time, when St. Christopher's Hospice opened in 1967, Miss Summers worked as a volunteer nurse on the wards for six years. When the Study Centre opened in 1973, she joined the staff as coordinator of studies and since then has developed an extensive educational program for the hospice staff and for professional visitors of all disciplines from all over the world.

Otto F. Thaler received the M.D. from the University of Rochester and completed his psychiatric residency there in 1955. He has been a member of the University of Rochester Medical School faculty since then and, in addition, is a graduate of the Downstate Psychoanalytic Institute. Currently he is professor of psychiatry and clinical director of an inpatient floor in the Psychiatry Department as well as director of inpatient services. Dr. Thaler coauthored *Clinical Supervision of the Psychiatric Resident* (1972).

About the Editors

Doris Teichler-Zallen graduated from Brooklyn College and received the Ph.D. in biology at Harvard University. She has taught at the University of Rochester, the Technion—Israel Institute of Technology, and is currently on the faculty of biology at Nazareth College of Rochester. Her laboratory research has involved studies of the genetics and biochemistry of photosynthesis. Dr. Teichler-Zallen's recent research in the field of human genetics, on the use of linkage analysis for the prenatal detection of gene disorders, is carried out in collaboration with the Pediatrics-Genetics Division of the University of Rochester School of Medicine where she is a research Fellow. She has produced several television programs on bioethics, based on the Nazareth College Science and Morality conferences.

Colleen D. Clements is an assistant professor of psychiatry and codirector of the Program in Psychiatric Ethics at the University of Rochester School of Medicine. One of her interests in the program concerns developing residency education in clinical ethics. In addition, she is a Fellow in the Genetics Division of the Pediatrics Department. She received the Ph.D. in philosophy from the University of Rochester, specializing in medical ethics, and is currently a National Science Foundation (EVIST) Fellow. Dr. Clements's publications include a medical genetics and ethics casebook and articles published in philosophy, psychiatry, and medical journals on ethics and psychiatry, medical sociology, medical ethics, and systems theory. She is currently producing a television series on issues in medical ethics.